109th Annual Meeting of the South Dakota Academy of Science

Proceedings of the South Dakota Academy of Science
Volume 103

Aberdeen, South Dakota, USA
19-20 April 2024

ISBN: 979-8-3313-0813-1

Printed from e-media with permission by:

Curran Associates, Inc.
57 Morehouse Lane
Red Hook, NY 12571

Some format issues inherent in the e-media version may also appear in this print version.

Printed with permission by Curran Associates, Inc. (2025)

For permission requests, please contact South Dakota Academy of Science at the address below.

South Dakota Academy of Science
308 Orchard Blvd.
Sawyer MI 49125

Phone: (605) 996 7670

southdakotaacademyofscience.wildapricot.org/

Additional copies of this publication are available from:

Curran Associates, Inc.
57 Morehouse Lane
Red Hook, NY 12571 USA
Phone: 845-758-0400
Fax: 845-758-2633
Email: curran@proceedings.com
Web: www.proceedings.com

Proceedings
of the
South Dakota Academy of Science

Volume 103
2024

Published by the South Dakota Academy of Science
Academy Founded 22 November, 1915
Academy Website: https://southdakotaacademyofscience.wildapricot.org

The Proceedings of the South Dakota Academy of Science

Since its beginning in 1915, the South Dakota Academy of Science has been a forum for the promotion of scientific teaching, research, and service in the state of South Dakota. The South Dakota Academy of Science meets each year for the purpose of conducting annual business, presenting awards, and facilitating the interchange of ideas, information, and results from scientists and students of science. The minutes of the annual meeting and the abstracts and full papers of presentations are published annually in the Proceedings of the South Dakota Academy of Science, which is uploaded to the Academy website (https://southdakotaacademyofscience.wildapricot.org). Although presentation at the annual meeting is strongly encouraged, each year a limited number of full papers may be considered for publication that have not been presented. The Instructions for Authors can be viewed at www.sdaos.org.

This volume contains the minutes of the annual meeting which was held at the Northern State University, April 19 and 20, 2024, and various reports. Additionally, it contains the full papers and abstracts of talks and posters presented at the annual meeting.

TABLE OF CONTENTS

Full Papers Submitted for Presentation at the 109th Annual Meeting of the South Dakota Academy of Science

Abstracts of Research Papers and Posters Submitted for Presentation at the 109th Annual Meeting of the South Dakota Academy of Science

Research Papers and Posters Presentd at the 109th Annual Meeting of the South Dakota Academy of Science Published by Title Only

Manuscript Reviewers

COMBINED MINUTES 109TH ANNUAL MEETING SOUTH DAKOTA ACADEMY OF SCIENCE 19 APRIL 2024 (HYBRID) AND 20 APRIL 2024 (IN PERSON) NORTHERN STATE UNIVERSITY

A. Welcome by President Tom Montoya
B. Agenda was approved.
C. Reports
1. President's report by Tom Montoya--Will occur at the 2025 Annual Meeting
2. Secretary's report by Donna Hazelwood

   The 2023 SDAS Annual Not-for-Profit and annual Fee were filed with the South Dakota State Treasurer. The 2023 SDAS Annual hybrid meeting was hosted by USD on 27 and 28 October 2023. Copies of individual certificates for the annual Poster and Paper Competitions were provided to the SDAS treasurer. The 2024 SDAS hybrid Annual Meeting Executive Council meeting will be 2:00 pm on Friday 19 April 2024. The date for the fall SDAS Executive Council meeting is TBA.

   Respectfully submitted,
   Donna Hazelwood, SDAS secretary
3. Treasurer's report by Jeffrey Palmer

   Treasurers report was distributed by hand. The current balance in checking account including support of Science Fairs at NSU and SDSU is $8656.23.
4. Proceedings Editor's report by Robert Tatina (1 January 2024 to 26 April 2024)

   On April 15, the following libraries were mailed a complimentary paper copy of the 2023 Proceedings: AC, BHSU, DSU, DWU, MMC, NSU, OLC, USF, NSU, SBC, SGU, SDSMT, SDSU (2 copies), SWC, UND, USD (2 copies), SD State Library, Am. Mus. Nat. Hist., U of WA and the Booth Fish Hatchery. The following indexing/ abstracting services were mailed complimentary paper copies or files of PDFs of the 2023 Proceedings: EBSCOhost, SAGE Publishing (Abstracts in Anthropology), Thompson Scientific (Zoological Record). Electronic files of the 2022 Proceedings were sent to Miner Solutions for publishing to the SDAoS website.

   For the 2023 Proceedings:

   No. of paper copies printed = 25 (2022: 26)
   No. of pages = 98 (2022: 74)
   No. of full papers published = 2 (2022: 2)
   No. of abstracts published = 12 (2022: 9)
   No. of titles only published = 1 (2022: 0)

No. paper copies distributed to libraries, abstracting services, individuals, etc.= 25
75 pages invoiced @ $20.00 per page = $1500.00 (Total paid = $1420.00)

Four calls for papers were sent out between January and April. 2024 Annual Meeting Program sent 12 April 2024 to presenters.

PROCEEDINGS LEDGER			Expenses	Income	
	1/1/24	Beginning Balance			3982.43
2664	3/15/24	Holmlund--formatting 2023 PSAS	710.00		3272.43
DEP	4/9/24	8 checks--page charges		1400. 00	7672.43
2665	4/12/24	MMP--print & mail 2023 PSAS	843.74		3828.69
	4/18/24	Ending Balance			3828.69

Submitted by Robert Tatina, Editor

5. Fellows Nomination Committee report by Bob Tatina-no report
6. Membership Committee report by Tim Mullican-no report
7. Social media report by Tim Mullican-no report
8. Publicity report by Dave Bergmann-no report
9. Webmaster's report by Tim Mullican

Following our Fall Annual Meeting in October of 2023, I had Shane Miner take down the online meeting registration page and the mail-in registration page on our website. We still had online annual dues payment set up on the web page. Then, this spring, I sent the First Call for Papers from Bob Tatina to Shane to post on the website. I also received a couple of emails from members who noted problems with the website, which I forwarded to Shane. It seemed that the turn-around time kept getting slower and slower. When it came time to set up the online payment for the 2024 Annual Meeting, the response was very slow. Finally, Bob Tatina graciously looked up Shane's phone number and called him. After Bob's conversation with him, I received the following email from Shane:

Tim,
I'm definitely having some issues with the site - but I was able to get the registration live now.
Bob - got the voicemails, and yes I'm the COO (not the CEO) of GBH in Boston and it is a big job (thanks on the congrats - I've been at GBH for nearly 10 years, I was the CTO previously working for COO and was promoted). I'm still working on the proceedings, but with the issues it isn't there yet.
And yes - I'd love to get out of the web hosting business with you all, but I definitely don't want to leave you in the lurch either. Happy to transfer all of the files/proceedings/etc to anyone if you

think you have someone who could build you something new. I actually think you could even just set up a SquareSpace site - and it could handle your registrations and membership all built in. The library would take a bit to set up - but I'm happy to share the files that I used to create it (which just make it into easy to use Excel/CSV docs).
If you do have anyone in mind to try and take things over - let me know, I'm happy to talk to them about ideas/options/etc.
Thanks, Shane

I had a conversation with Paula Mazzer, and she said that she is a member of an organization that uses Wild Apricot for their web hosting (https://freetrial.wildapricot.com/). They advertise that they provide web hosting and membership services for over 26,000 organizations. I am a member of the American Society of Mammalogists, who have members worldwide, and they also use Wild Apricot for sending out membership dues reminders and payment processing. It seems to work very well. Paula said that the cost is like what we have been paying Shane, but it sounds like it would do even more for us. Wild Apricot offers a free 60-day trial. I would be willing to set up a website using their free trial and let the Executive Committee take a look at it to see what they think. I have set up webpages myself many moons ago using Wordpress, Frontpage as well as MS Word. It looks like this would be even easier and more reliable. If one of us sets it up, we would be able to make changes much more quickly than in the past. I will be finishing up classes in just a few weeks and will have time to play around with it. I think I could even set up a page with links to the Proceedings. Please let me know if you would like me to go ahead with this. It sounds like it would be a fun project.

Submitted by Tim Mullican, Webmaster

After the Executive Council agreed to have Tim move forward with a free 60-day Wild Apricot trial, this discussion followed: Paul Johnson suggested, and Tom Montoya expanded on the idea that the EC consider a scheme for allowing transitions in the future bolstered by redundancies in the system as migration to the Wild Apricot System occur. Jeff Palmer questioned if all of the Proceedings would be, or would be made, searchable. He also inquired about migrating PayPal. Paul raised the question about searchable digitized documents for each of the Proceedings. This inquiry will be posed to Bob. Paul also inquired about whether the EC would support having the Proceedings be hosted by SDSU Briggs Library Open Prairie. Discussion will continue after the status of original digital files, especially for the earlier Proceedings, have been examined. Paul asked if the EC would agree to have the Proceedings published online, open-access, by Open Prairie. Discussion will continue at the fall EC meeting.

Paul had a question for Tim about whether a digital file exists for the SDAS banner on the SDAOS website. Discussion will continue.

D. Elections: The slate of SDAS Executive Council nominees for 2024-2026 Executive Council was elected by members present at the 2024 SDAS Annual Business Meeting hosted by NSU.

Position	Name	Institution	Dates
Member-at-Large	Arvid Boe	SDSU	2024-2026
Member-at-Large	Bruce Cuevas	USD	2024-2026
Member-at-Large	Paul Johnson	SDSU	2024-2026
Member-at-Large	Xu Lan	SDSU	2024-2026
Member-at-Large	Abigail Mertens	SDSU	2024-2026
Member-at-Large	Daniel Owens	SDSM&T	2024-2026
Member-at-Large	Diane Roeder	AU	2024-2026
Member-at-Large	Rajesh Shende	SDSM&T	2024-2026
Member-at-Large	Purushotham Tukkaraja	SDSM&T	2024-2026
Member-at-Large	Chao Zhang	MMU	2024-2026

E. Old/Continuing Business
1. Web host; Tim Mullican: discussion will continue
2. Continuing Discussion of Bob's proposal to overhaul the Membership Committee and requests for individuals to step forward and champion topics in their interest and expertise. To be continued.
3. Discuss funding by the SDAS for the new researcher award championed by Mark Gabel, currently in the second year of three years funding by an anonymous donor. To be continued.

F. Audit; Mark Gabel and Donna Hazelwood: All documents in order. Mark congratulated Jeff on another year of excellent easy to follow record keeping.

G. Upcoming meetings:
1. Fall 2024 EC Zoom Meeting TBA
2. 110th Annual Meeting hosted by SDSM&T 2025: Contact Tom Montoya; Date and location TBA
3. 111th Annual Meeting hosted by USF 2026: Contact
4. 112th Annual Meeting hosted by DWU: Contact.
5. 113th Annual Meeting hosted by SDSU 202
6. 114th Annual Meeting hosted by MMU 202
7. 115th Annual Meeting tentatively hosted by SDAS EC Chamberlain (2024)

H. Resolutions by Donna Hazelwood, Resolutions Committee

The Members of the South Dakota Academy of Science sincerely and gratefully thank the following:

1. Thank you to the Local Planning and Organizing Committee Co-Chairs Eric Pulis and George Nora, who through dedicated teamwork hosted the SDAS Annual meeting on the NSU campus, while also making available Zoom connections where feasible. The SDAS is sincerely grateful.
2. Sincere special heartfelt thanks to long-standing well organized and timely Proceedings Editor Bob Tatina for his continued excellence and dedication organizing and compiling the schedule for papers and posters and for his outstanding dedication as editor of the Proceedings of the SDAS. The SDAS also thanks Bob for chairing the Fellows Nomination Committee;
3. We sincerely thank President Tom Montoya for masterfully guiding the SDAS through the first year of his presidency with well thought out direction and considerable expertise focused on the members of the SDAS.
4. Sincere gratitude and recognition to Past President Tim Mullican (2021- 2023) for his tireless dedicated active and pioneering service to the SDAS through changing times, and for fostering community and science throughout the year. Always going above and beyond. In addition, sincere thanks to Tim for also serving as competent and dedicated Webmaster, Chair of Membership and chair of Social Media committees and for his upcoming planning got a pioneering trial of the web host Wild Apricot;
5. Special thanks to Donna Hazelwood for continuing to serve as Secretary;
6. Special thanks to the dedicated, well organized and meticulous Jeff Palmer, Treasurer;
7. Thanks to Tim Mullican for doing dedicated double duty and continuing to serve as Web Master;
8. We thank Fellows Committee Chair Bob Tatina for service to the SDAS Fellows and membership and for inviting SDAS Fellows to nominate members;
9. Thank you to all members of the SDAS Executive Council for ongoing, innovative, valuable and generous service to the SDAS
10. Thank you to all members of the SDAS for ongoing, renewed and new support of the SDAS
11. We thank the Resolution Committee: Donna Hazelwood;
12. We thank Mark Gabel and Donna Hazelwood for service on the Audit Committee,
13. We thank members of the Poster Judging Committee, Adam Sturlaugon, Eric Pulis and George Nora.

14. We congratulate the awardees of Poster Judging 2024
    Poster award
        Jessica Splichal. Science & Mathematics, NSU "TRANSCRIPTION ANALYSIS OF MIRNA'S WITHIN CD34+ HEMATOPOIETIC STEM CELLS"
15. We thank oral paper judges, Tom Montoya and Adam Sturlaugson.
16. We congratulate awardees of Oral Presentation Judging 2024.
    Undergraduate student Oral Presentation award
        Zoe Boughton and Eric Pulis. Science & Mathematics, NSU. "MOLECULAR SCREENING AND IDENTIFICATION OF BLOOD PARASITES FROM SOME UPLAND GAMEBIRDS (PHASIANIDAE)"
    Graduate student Oral Presentation awards
        Sanam Parajuli, Bibek Adhikari, and Madhav P. Nepal. Biology & Microbiology, SDSU. "GENOME-WIDE IDENTIFICATION OF MADS-BOX TRANSCRIPTION FACTOR GENES IN RED MULBERRY (*MORUS RUBRA* L.)"
        Bharathkiran Maddipudi, Khang Huynh, Anuradha Shende, and Rajesh Shende. Chemical & Biological Engineering, SDSM&T. "CATALYTIC HYDROTHERMAL LIQUEFACTION OF AGRICULTURE RESIDUE FOR VALUE-ADDED PRODUCTS"
17. We welcome the following new members of the SDAS Executive Council: Abigail Mertens, Daniel Owens, Diane Roeder, Rajesh Shende and Purushotham Tukkaraja.
18. We thank Keynote Speaker Manu Manu, University of North Dakota, for his Keynote Address: "How do Stem Cells Choose Their Fate?"
19. We thank everyone for participating in the 109th SDAS Annual Meeting hosted by NSU and for mentoring students, encouraging poster and paper presentations and for introducing students in person to the SDAS via the Annual Meeting.

Respectfully submitted,
Donna Hazelwood, SDAS Secretary

TREASURER'S REPORT

South Dakota Academy of Science
Treasurer's Report 01 Jan 2024 - 31 Dec 2024

01-Jan-24	Checking Account Balance	$7,826.23			
	Certificate of Deposit	$9,178.69			
	Petty Cash	$0.00			
	Total Assets	$17,004.92			

	Income	Expense	Date	Check #	Payee
Meeting Registrations and Dues	110.00				
Northern Science Fair		240.00	21-Mar-24	1498-1505	Science Fair Awards
Eastern SD Science & Engineering Fair		240.00	18-Mar-24	1506	Science Fair Awards
Transfer from PayPal	1200.00		10-Apr-24		
Meeting Registration Forms		22.00	16-April-24	1507	DSU Print Shop
Meeting Registrations and Dues	1150.00		29-Apr-24		
Grad Oral 1st		300.00	13-May-24	1508	Sanan Parajuli
Undergrad Oral 1st		300.00	13-May-24	1509	Zoe Boughton
Grad Oral 2nd		200.00	13-May-24	1510	B. Maddipudi
Poster 1st		300.00	22-May-24	1511	Jessica Splichal
Meeting Expenses		604.40	24-May-24	1512	NSU
Transfer from PayPal	1009.97		29-May-24		
Domestic Nonprofit Filing Fee		10.00	14-Jun-24	1206	Secretary of State, SD
NAAS Affiliation Dues		100.00	13-Jul-24	1513	NAAS
Totals	**3469.97**	**2316.40**			

31-Dec-23	Checking Account Balance	$8,979.80			
	Certificate of Deposit	$9,178.69			
	Petty Cash	$0.00			
	Total Assets	$18,158.49			

Respectfully Submitted,
Jeffrey S. Palmer

2024 PROCEEDINGS EDITOR'S REPORT

1 January 2024 to 26 April 2024

A. On April 15, the following libraries were mailed a complimentary paper copy of the 2023 Proceedings: AC, BHSU, DSU, DWU, MMC, NSU, OLC, USF, NSU, SBC, SGU, SDSMT, SDSU (2 copies), SWC, UND, USD (2 copies), SD State Library, Am. Mus. Nat. Hist., U of WA and the Booth Fish Hatchery.
B. The following indexing/ abstracting services were mailed complimentary paper copies or files of PDFs of the 2023 Proceedings: EBSCOhost, SAGE Publishing (Abstracts in Anthropology), Thompson Scientific (Zoological Record)
C. Electronic files of the 2022 Proceedings were sent to Miner Solutions for publishing to the SDAoS website.
D. For the 2023 Proceedings:
    1. No. of paper copies printed = 25 (2022: 26)
    2. No. of pages = 98 (2022: 74)
    3. No. of full papers published = 2 (2022: 2)
    4. No. of abstracts published = 12 (2022: 9)
    5. No. of titles only published = 1 (2022: 0)
    6. No. paper copies distributed to libraries, abstracting services, individuals, etc.= 25
    7. 75 pages invoiced @ $20.00 per page = $1500.00 (Total paid = $1420.00)
E. Four calls for papers were sent out between January and April.
F. 2024 Annual Meeting Program sent 12 April 2024 to presenters.
G. Proceedings Ledger

			Expenses	Income	
	1/1/24	**Beginning Balance**			**3982.43**
2664	3/15/24	Holmlund—formatting 2023 PSAS	710.00		3272.43
DEP	4/9/24	8 checks—page charges		1400. 00	7672.43
2665	4/12/24	MMP—print & mail 2023 PSAS	843.74		3828.69
	4/18/24	**Ending Balance**			**3828.69**

Submitted by Robert Tatina, Editor

2024-2026 EXECUTIVE COUNCIL

President	Thomas Montoya, SDSM&T, Electrical Engineering
2023-2025	394-2459, Thomas.Montoya@sdsmt.edu
Vice President	Paula Mazzer. DWU, Chemistry.
2023-2025	995-2899, Paula.Mazzer@dwu.edu
Past President	Tim Mullican, DWU, biology
2023-2025	995-2715, Tim.Mullican@dwu.edu
Secretary	Donna Hazelwood, DSU, Biology, (retired)
	256-5187, donna.hazelwood@dsu.edu
Treasurer	Jeffrey Palmer (retired), DSU, Mathematics
	256-5190, Jeff.Palmer@dsu.edu
Proceedings Editor	Robert Tatina, DWU (retired)
	269-426-3744 Robert.Tatina@dwu.edu
Chair Fellows Nominations Committee	Robert Tatina, DWU (retired)
	269-426-3744 Robert.Tatina@dwu.edu
Co-Chair Membership Committee	Robert Tatina, DWU (retired)
	269-426-3744 Robert.Tatina@dwu.edu
Co-Chair Membership Committee	Open
Co-Chair Membership Committee	Open
Co-Chair Social Media Committee	Tim Mullican, DWU, biology
	995-2715, Tim.Mullican@dwu.edu
Co-Chair Social Media Committee	Grad Student Open
Publicity	Dave Bergmann, BHSU, Biology,
	642-6465 Dave.Bergmann@bhsu.edu
Publicity	Robert Tatina, DWU biology (retired)
	269-426-3744; Robert.Tatina@dwu.edu
Webmaster	Tim Mullican, DWU, biology
	995-2715, Tim.Mullican@dwu.edu
Member-at-Large	Mark Gabel, BHSU biology (retired)
2023-2025	642-6251, Mark.Gabel@bhsu.edu
Member-at-Large	Abigail Martens, SDSU, Entomology, Insect Biodiversity Lab
2023-2025	688-4593, abigail.martens@sdstate.edu
Member-at-Large	Geore Nora, NSU,Chemistry
2023-2025	626-2471, george.nora@northern.edu
Member-at-Large	Eric Pulis, NSU, Biology
2023-2025	626-2600, eric.pulis@northern.edu

Member-at-Large 2023-2025	Adam Sturlaugson, USF, Chemistry 331-6757, Adam.Sturlaugson@usiouxfalls.edu
Member-at-Large 2023-2025	Open
Member-at-Large 2023-2025	Open
Member-at-Large 2023-2025	Open
Member-at-Large 2024-2026	Arvid Boe (retired), SDSU, Plant Science 688-4759, Arvid.Boe@sdstate.edu
Member-at-Large 2024-2026	Bruce Cuevas, USD, Basic Medical Sciences 658-6330, Bruce.Cuevas@usd.edu
Member-at-Large 2024-2026	Paul Johnson (retired), SDSU, Entomology 688-4438, conoderus@gmail.com
Member-at-Large 2024-2026	Daniel Owens, Chemistry, Biology and Health Sciences Daniel.Owens@sdsmt.edu, (605) 394-1238
Member at Large, 2024-2026	Diane Roeder, AU, Biology 274-4737, Diane.Roeder@augie.edu
Member-at-Large 2024-2026	Rajesh Shende (Chemical and Biological Engineering Dept) Rajesh.Shende@sdsmt.edu, (605) 394-1231
Member-at-Large 2024-2026	Purushotham Tukkaraja (Mining Engineering and Management Purushotham.Tukkaraja@sdsmt.edu, (605) 394-1267
Member-at-Large 2024-2026	Xu Lan, SDSU, Department of Natural Resource Management 688-4564. Lan.Xu@sdstate.edu
Member-at-Large 2024-2026	Chao Zhang, MMU, Physics 668-1588, chao.zhang@mountmarty.edu
Graduate Student Member-at-Large 2024-2026	Open

FELLOWS OF THE SOUTH DAKOTA ACADEMY OF SCIENCE 1998-2023

1988	Kenneth F. Higgins (SDSU)
1999	The late Chuck Estee (USD)
2001	Carroll Hanten
2001	Emil F. Knapp (AU)
2003	The late Clyde Brashier (DSU)
2003	The late Milton Hanson (AU)
2003	The late Sr. Laeticia Kiltzer (MMC)
2003	The late Robert Looyenga (SDSM&T)
2003	Arlen Viste (AU)
2003	The late Everett White (SDSU)
2004	Perry H. Rahn (SDSM&T)
2004	Robert Stoner (USD)
2011	Maureen Diggins-Hutcheson (AU)
2011	Gary Earl (AU)
2011	The late H.L. Hutcheson (SDSU)
2011	Wally Klawiter Jr. (USF)
2012	Nels Granholm (SDSU)
2012	The late Audrey Gabel)BHSU)
2012	Mark Gabel (BSU)
2012	Robert Tatina (DWU)
2015	Arvid Boe (SDSU)
2015	R. Roy Kintner (AU)
2015	The late Gary Larson (SDSU)
2015	The late Norm Miller (SDSM&T)
2015	David Parris (NJ Museum)
2016	Donna Hazelwood (DSU)
2016	Cliff Lewis (USF))
2016	James Martin (SDSM&T)
2016	R. Neil Reese (SDSU)
2017	Dave Bergmann (BHSU)
2017	Andrew Detwiler (SDSM&T)
2019	Nels Troelstrup (SDSU)
2020	Larry Stetler (SDSM&T)
2020	Duane Weisshaar (AU)
2021	Mile Koppang (USD)
2021	Paul Johnson (SDSU)
2022	Jeffrey Palmer (DSU)
2022	David Swanson (USD)

HONOR ROLL OF SOUTH DAKOTA ACADEMY OF SCIENCE PAST PRESIDENTS

prelim	1914-1915	†Hilton Ira Jones, Chemistry, Dakota Wesleyan University
1st	1915-1916	†Hinton Ira Jones, Chemistry, Dakota Wesleyan University
2nd	1916-1917	†A.N. Cook, Chemistry, University of South Dakota
3rd	1917-1918	†A.N. Hume, Agronomy, State College of Agridulture & Mechanic Arts
4th	1918-1919	†J. Gladden. Hutton, Soils, State College of Agriculture & Mechanic Arts
5th	1919-1920	†Freeman Ward, Geology, University of South Dakota
6th	1920-1921	†Charles W. Cuno, Chemistry, Yankton College
7th	1921-1922	†Doane Robinson, Historian, Pierre
8th	1922-1923	†A.N. Hume, Agronomy, State College of Agriculture & Mechanic Arts
9th	1923-1924	†Arthur M. Pardee, Chemistry, University of South Dakota
10th	1924-1925	†L.A. Stout, Mathematics, Dakota Wesleyan University
11th	1925-1926	†J. Gladden Hutton, Soils, State College of Agriculture & Mechanic Arts
12th	1926-1927	†A.P. Larrabee, Biology, Yankton College
13th	1928	†Homer Hill, Science, Huron College
14th	1928-1929	†H.C. Severin, Entomology, State College of Agriculture & Mechanic Arts
15th	1929-1930	†E.P. Rothrock, Geology, University of South Dakota
16th	1930-1931	†V.A. Lowry, Science, Eastern State Teacher's College
17th	1931-1932	†B. B. Brackett, Communications, University of South Dakota
18th	1932-1933	†George Gilbertson, Entomology, State College of Agriculture & Mechanic Arts
19th	1933-1934	†A.V. Arlton, Biology, Dakota Wesleyan University
20th	1934-1935	†E.P. Churchill, Biology, University of South Dakota
21st	1935-1936	†William H. Powers, Librarian, State College of Agriculture & Mechanic Arts
22nd	1936-1937	†Greg M. Evans, Chemistry, Yankton College
23rd	1937-1938	†Charles A. Hunter, Bacteriology, University of South Dakota
24th	1938-1939	†Ward L. Miller, Botany, State College of Agriculture & Mechanic Arts
25th	1939-1940	†Walter V. Scaright, Geology, University of South Dakota
26th	1940-1941	†A.L. Moxon, Chemistry, State College of Agriculture & Mechanic Arts
27th	1941-1942	†Henry Lowsma, Science, General Beadle State College
28th	1942-1943	†Edwin H. Shaw, Jr., Biochemistry, University of South Dakota
29th	1943-1944	†Orin M. Lofthus, Biology, Augustana College
30th	1944-1945	†Lester S. Guss, Chemistry, South Dakota State College
31st	1945-1946	†Raymond J. Greb, Biology, Huron College
32nd	1946-1947	†F.L. Bennett, Botany, Black Hills Teachers College
33rd	1947-1948	†A.L. Haines, Chemistry, University of South Dakota
34th	1948-1949	†V.S. Webster, Chemistry, South Dakota State College
35th	1949-1950	†John M. Froemke, Chemistry, Augustana College
36th	1950-1951	†William E. Ekeman, Mathematics, University of South Dakota
37th	1951-1952	†E R. Binnewies, Chemistry, South Dakota State University
38th	1952-1953	John M. Winter, Botany, University of South Dakota
39th	1953-1954	H.R. Fossler, Psychology, University of South Dakota
40th	1954-1955	†Harlan L. Klug, Chemistry, South Dakota State University
41st	1955-1956	†Frank W. Jobes, Biology, Yankton College
42nd	1956-1957	†Charles R. Estee, Chemistry, University of South Dakota
43rd	1957-1958	†John Willard, Chemistry, South Dakota School of Mines & Technology
44th	1958-1959	Shapley W. Howell, Science, Yankton College
45th	1959-1960	†V.R. Nelson, Physics and Aeronautics, Augustana College
46th	1960-1961	†Ernest J. Hugghins, Parasitology, South Dakota State University

47th	1961-1962	†Charles M. Vaughn, Science, University of South Dakota
48th	1962-1963	†Morton Green, Biology, South Dakota School of Mines & Technology
49th	1963-1964	†S. G. Froiland, Biology, Augustana College
50th	1964-1965	†George P. Scott, Chemistry, University of South Dakota
51st	1965-1966	†Marvin Hanson, Mathematics, Huron College
52nd	1966-1967	†Robert Sandvig, Chemistry, South Dakota School of Mines & Technology
53rd	1967-1968	†Walter Morgan, Biology, South Dakota State University
54th	1968-1969	†Charles Sidlo, Physics, Sioux Falls College
55th	1969-1970	†C.L. Hills, Biology, Dakota Wesleyan University
56th	1970-1971	R. Roy Kintner*, Chemistry, Augustana College
57th	1971-1972	†Raymond D. Dillon, Zoology, University of South Dakota
58th	1972-1973	†Veronica Fasbender, Biology, Mount Marty College
59th	1973-1974	David Ballew, Mathematics, South Dakota School of Mines & Technology
60th	1974-1975	†Jack K. Saunders, Biology, Northern State University
61st	1975-1976	†Harold Foss, Natural Science, Augustana College
62nd	1976-1977	Waldemar F. Klawiter, Jr.*, Mathematics and Physics, Sioux Falls College
63rd	1977-1978	†James C. Schmulbach, Biology, University of South Dakota
64th	1978-1979	James A. Fries, Chemistry, Northern State University
65th	1979-1980	†John Landegent, Biology, Sioux Falls High School
66th	1980-1981	†Norman E. Miller, Chemistry, University of South Dakota
67th	1981-1982	Harry G. Hecht, Chemistry, South Dakota State University
68th	1982-1983	Theodore Van Bruggen, Botany, University of South Dakota
69th	1983-1984	L. Cliff Lewis*, Chemistry, Sioux Falls College
70th	1984-1985	Maureen Diggins-Hutcheson*, Reproductive Biology, Augustana College
71st	1985-1986	Don Hopkins, Physics, South Dakota School of Mines & Technology
72nd	1986-1987	Stephen R. Metzner, Science, University of South Dakota
73rd	1987-1988	†Marvin Selnes, Science, Sioux Falls School District
74th	1988-1989	Warren Hein, Physics, NSU/SDSU
75th	1989-1990	James Martin*, Paleontology, South Dakota School of Mines & Technology
76th	1990-1991	Samuel Gingerich, Chemistry, Northern State University
77th	1991-1992	Mark Gabel*, Botany, Black Hills State University
78th	1992-1993	Nels Granholm*, Biology and Global Studies, South Dakota State University
79th	1993-1994	Keith Perkins, Biology, Sioux Falls College
80th	1994-1995	Arvid Boe*, Agronomy, South Dakota State University
81st	1995-1996	Tim Sorenson, Mathematics, Augustana College
82nd	1996-1997	John Thomas, Biology, University of South Dakota
83rd	1997-1998	Sharon Clay, Plant Science, South Dakota State University
84th	1998-1999	Royce Engstrom, Chemistry, University of South Dakota
85th	1999-2000	Neil Reese*, Botany, South Dakota State University
86th	2000-2001	Lenore Koczon, Chemistry, Northern State University
87th	2001-2002	Charles Lamb, Biology, Black Hills State University
88th	2002-2003	†Steve McDowell, Chemistry, South Dakota School of Mines & Technology
89th	2003-2004	Miles Koppang*, Chemistry, University of South Dakota
90th	2004-2005	Andy Detwiler*, Atmospheric Sciences, S D School of Mines & Technology
91st	2005-2006	Robert Tatina*, Biology, Dakota Wesleyan University
92nd	2006-2007	James Sorenson, Biology, Mount Marty College
93rd	2007-2008	Michael Wanous, Biology, Augustana College
94th	2008-2009	Nels H. Troelstrup, Jr., Biology, South Dakota State University
95th	2009-2010	Krisma DeWitt, Chemistry, Mount Marty College
96th	2010-2011	Dave Bergmann*, Biology, Black Hills State University
97th	2011-2012	Krisma DeWitt, Chemistry, Mount Marty College

98th	2012-2013	†Gary Larson*, Biology, South Dakota State University
99th	2013-2014	†Gary Larson*, Biology, South Dakota State University
100th	2014-2015	Chun Wu, Chemistry, Mount Marty College
101st	2015-2016	Steven Matzner, Biology, Augustana University
102nd	2016-2017	George Mwangi, Chemistry, University of Sioux Falls
103rd	2017-2018	Vicki Geiser, Northeast Community College
104th	2018-2020	Vicki Geiser, Freeman Regional Hospital
105th	2020-2023	Tim Mullican, Dakota Wesleyan University

*Fellow of the SD Academy of Science
†Deceased

MEMBERSHIP

2023 Membership List

Abdulallah Alomair
Arvid Boe
Lexa Burtzlaff
Abhinav Chandel
Robert Cihak
Bruce Cuevas
Kaesh Davis
Andrew Detwiler
Mark Dixon
Kaleb Dobbs
Dale Droge
Mark Gabel
Kathleen Gibson
Lin Guo
Mark Hanhardt
Stevie Holmes
Khushboo Jain
Meghann Jarchow
Ranjeet John
Paul J. Johnson
Michaela Jones
Venkatesh Kolluru
Madison Kovar
Aaron Madden
Chelsi Marolf
Abigail Martens

2024 Membership List

Bibek Adhikari
Abdulallah Alomair
Arvid Boe
Zoe Boughton
Sage Bultje
Summer Carlson
Bruce Cuevas
Amy Dolan
Megan Fastenau
Mark Gabel
Oxana Gorbatenko
Mark Hanhardt
Khang Huynh
Paul Johnson
John Long
Bharathkiran Maddipudi
Michael May
Daryl E. Mergen
Zach Mergen
Thomas P Montoya
Tim Mullican
Emma Nelson
Madhav Nepal
George Nora
Jeffrey Palmer
Sanam Parajuli
L. Brian Patrick
Eric Pulis
Neil Reese
Diane Roeder
Jessica Splichal
Samantha St. Clair
Adam Sturlaugson
Robert Tatina
Jessica Warns
Henry Wei

THE BY-LAWS OF THE SOUTH DAKOTA ACADEMY OF SCIENCE 2020

ARTICLE I
Members

SECTION 1. The membership of the Academy shall consist of members, associate members, honorary members, and corporate members. The right to vote and hold office shall be reserved for members and associate members.

SECTION 2. Members shall be those persons interested in scientific work and research with the exception of those persons specified in Article I, Sections 3, 4, and 5. Any member or applicant for membership who at one time shall contribute to the funds of the Academy ten times the annual dues or any individual who has been a member of the South Dakota Academy of Science for twenty-five years and who has reached the age of sixty-five years, may be nominated to life membership. Life members shall be exempt from payment of all membership dues and annual meeting fees thereafter. Academy Fellows (Article I, Section 7) are considered life members and shall be exempt from membership dues.-

SECTION 3. Associate members shall be undergraduate/graduate students interested in scientific work.

SECTION 4. Honorary members. Any person who has rendered conspicuous service in the advancement of science may be nominated by a member to honorary membership of the Academy. Honorary members shall be exempt from all membership dues and annual meeting fees.

SECTION 5. Corporate members shall consist of such organizations and business concerns which wish to affiliate with the South Dakota Academy of Science. Such entities may be nominated by a member to membership of the Academy and, if elected, shall pay annual dues as determined and published by the Academy.

SECTION 6. Any individual may become a Member of the Academy upon payment of annual dues. Any undergraduate or graduate student may become an Associate Member upon payment of annual dues. Honorary Members and Life Members may be elected by a vote of members at an Annual Meeting upon recommendation of the Membership Committee. A majority of those voting will be required for election. A member may be dismissed from the organization if they have exhibited any of the following: fraud, conflict of interest, improper personal conduct, lack of fitness to serve, or failure to perform assigned duties. The vote to remove a member shall take place at a regularly called meeting of the Executive Council, and shall be approved by a vote of simple majority of the Executive Council.

SECTION 7. Academy Fellows. A member of the South Dakota Academy of Science may be designated as a Fellow, an honor afforded to those members whose years of service to the Academy and the promotion of science in the state and nation is meritorious. Nominations for Fellows will be solicited annually from the membership. Nominees must be active members of the Academy for a total of at least 10 years. The nominee will have made outstanding contributions in the area of specialization whether in research, teaching, extension, service, or private service activities. The nomination will include a letter of recommendation about the nominee including degrees received, professional positions held, membership in honorary academic societies, honors and awards received since baccalaureate degree, service to SDAS in appointed and elected positions, and professional achievements in science (teaching, extension, investigative, service, leadership). Review of nominations and designation of the title Fellow will be initially conducted by the Executive Council. When the membership of active Fellows reaches a critical mass (more than 10 members), the review and designation of Fellow will be conducted by the active group of Fellows. If the membership of active Fellows falls below the critical mass, then the Executive Council will assist with the selection of Fellows.

ARTICLE II
Officers

SECTION 1. The officers of the Academy shall consist of a President, a Vice-President, a Past- President, a Secretary, a Treasurer, an Editor of the Proceedings of the South Dakota Academy of Science, and four to eight Members-at-Large.

SECTION 2. The Vice-President shall be elected every two years by a majority vote of those members voting or as specified in Article II Section 5, and will serve for three two-year sequential terms, Vice-President, President, and Past-President. Members-at-Large are to be elected from the membership, including associate members to serve for two-year terms except that approximately half shall be elected each year. At the time of the initial election of these members, half will be elected for a one-year term.

SECTION 3. It shall be one of the duties of the President to prepare an address which shall be delivered before the Academy at the Annual Meeting. In a presidential transition year, the address will be given by the incoming President.

SECTION 4. The President shall present to the members at the Annual Meeting such actions of the Executive Council as are contained in Article III, Section 2 and those subject to ratification by the membership.

SECTION 5. Upon death or resignation of the President, the Vice-President shall immediately assume the office of the President; upon the death or resignation of the Vice-President, the vacant office shall be filled by the Executive Council.

ARTICLE III
Executive Council

SECTION 1. The Executive Council of the South Dakota Academy of Science shall consist of the officers of the Academy as specified in Article II, Section. The Executive Council members in attendance at scheduled meetings shall constitute a quorum.

SECTION 2. The Executive Council shall be entrusted the management of the affairs of the Academy during the intervals between regular meetings. No extraordinary act of the Executive Council shall remain in force beyond the next Annual Meeting without specific ratification by a majority of the members. The President may call special meetings of the Executive Council and shall call such a meeting upon written request by three members. The members of the Executive Council shall constitute the membership of the Board of Directors of the South Dakota Academy of Science.

SECTION 3. The Secretary, the Treasurer, and the Editor of the *Proceedings* shall be selected by the Executive Council. Term length of those positions shall be negotiated between the individuals selected and the Executive Council.

ARTICLE IV
Committees and Appointed Positions

SECTION 1. The standing committees of the Academy shall be a Committee on Publications, a Committee on Membership, a Resolutions Committee, a Nominating Committee, and an Auditing Committee. The President may appoint ad hoc committees as needed. Appointed Positions shall include, but not be limited to, Webmaster, Fellows Nomination Chair, and Social Media Chair.

SECTION 2. The Committee on Publications shall consist of the President, the Secretary, and the Editor. In addition, this committee may call upon one or more members to serve in any specific case. This committee shall be empowered to specify the conditions for the publication of papers in the Proceedings of the South Dakota Academy of Science. The Editor shall serve as chairperson of this committee.

SECTION 3. The Committee on Membership shall consist of at least two members who shall attempt to diversify and increase the membership. They shall be chosen by the Executive Council. The Membership Committee shall conduct an annual membership campaign among all educational, service and research institutions of the state. This campaign will include the distribution of informational brochures and membership applications. It is the duty of the Membership Committee to receive nominations for Honorary and Life Members and to make recommendations to the membership at the Annual Meeting.

SECTION 4. The Resolutions Committee shall consist of at least two members to be appointed annually by the President. The chairperson shall be appointed from the committee by the President. The Resolutions Committee is charged with the responsibility of presenting any special policies, recommendations, or recognitions to the membership of the Academy at the Annual Meeting for a vote of adoption.

SECTION 5. The Nominating Committee shall consist of at least two members chosen by the Executive Council. The chairperson shall be appointed from the committee by the President. It shall be the duty of this committee to recommend candidates for the several offices and for elected committees. This slate of officers shall be presented to the members and be voted on at the next Annual Meeting. Every two years this list shall present the names of two nominees for the office of Vice-President, one of whom shall be elected to the office. Nominations for candidates other than those presented by the nominating committee may be made by any member in accordance with Robert's Rules of Order.
SECTION 6. The President shall appoint the Academy Representative to the Executive Council of the American Association for the Advancement of Science and to the Academy Conference.

SECTION 7. The President shall appoint annually an Auditing Committee of at least two members ~~which~~ **who** shall audit the books of the Treasurer prior to the Annual Meeting and report the results of this audit at the business meeting.

SECTION 8. The Fellows Committee shall address issues of interest to Fellows of the Academy. The Fellows Committee shall consist of all active and interested Fellows.

SECTION 9. The Webmaster and Social Media Chair may appoint committee members to their committees as necessary.

ARTICLE V
Meetings

SECTION 1. The Annual Meeting of the Academy shall be held at such time and place as the Executive Council may designate. The Local Arrangements Committee shall decide if adverse conditions require cancellation of the annual meeting. Special Meetings of the Academy may be called by the Executive Council and shall be called upon the written request of fifteen members.

SECTION 2. No meeting of the Academy shall be held without notice sent by the Secretary or Editor to all members at least ten days before the date of the proposed meeting.
SECTION 3. The members present at an Annual or Special Meeting shall constitute a quorum of the Academy.

SECTION 4. If an Annual Meeting is cancelled due to weather or other unforeseen circumstances:
1) Submitted papers and abstracts may be published in the forthcoming volume of the *Proceedings.*
2) The term of officers of the Academy shall be extended as necessary to the next Annual Meeting.
3) Registration fees for the meeting will be refunded to those who paid advance registration.
4) Membership dues will not be refunded.

SECTION 5. At the time of the Annual Meeting, the Annual Banquet and Academy Lecture will be held. This Annual Meeting will ordinarily be held at one of the colleges or universities in the state of South Dakota or at an alternative site with suitable facilities. The host institution shall be determined by the Executive Council. The host institution will select a Local Arrangements Chairperson and the speaker who will deliver the Annual Academy Lecture. The host institution for the following year must send a representative from that institution to the Executive Council meeting and Annual Meeting in the year preceding the hosting of the meeting.

ARTICLE VI
Publications

SECTION 1. The regular publications of the Academy shall include the *Proceedings of the South Dakota Academy of Science*, which will contain such papers and abstracts by Members and Associate Members of the Academy, as are deemed suitable by the Committee on Publications, as stated in Article IV, Section 2.

ARTICLE VII
Business

SECTION 1. The following shall be the regular order of business for the Annual Meeting of the Academy:
1. Call to order and report by the President, outgoing and incoming, when applicable
2. Report of the Secretary
3. Report of the Treasurer
4. Reports of Officers
5. Reports of Standing Committees
6. Reports of Ad Hoc Committees
7. Unfinished Business
8. New Business
9. Election of Officers
10. Appointment of Special Committees
11. Adjournment

The program may be interpolated in the order of business according to convenience.

SECTION 2. Electronic voting either by the Executive Council or the Membership will be allowable based upon the recommendation of the President for electronic voting by the Executive Council or the approval of the Executive Council for electronic voting by the membership.

ARTICLE VIII
Fiscal Year

SECTION 1. The fiscal year of the Academy shall extend from January 1 through December 31.

ARTICLE IX
Dues and Assessments

SECTION 1. Each member shall pay annual dues as determined by the Academy.

SECTION 2. Membership extends from January 1 until the next scheduled Annual Meeting of the following year. Annual dues are not pro-rated.

ARTICLE X
Duties of the Secretary, the Treasurer, and the Editor

SECTION 1. The Secretary and/or members of the Executive Council shall conduct correspondence relating to the Academy. (Note for example: The *Proceedings* Editor conducts all correspondence regarding the *Proceedings*.) The Secretary shall take and publish minutes of the annual meeting in the *Proceedings of the South Dakota Academy of Science*. The Secretary shall keep, update and distribute a list indicating the rotation for locations of annual meetings. The Secretary shall organize the fall Executive Council meeting including date, location and food as necessary. The agenda shall include known vacancies on the SDAS Executive Council as advance notice for elections for the Business Meeting at the Annual Meeting. The Secretary shall take the minutes of the fall Executive Council meeting and publish those minutes to the Executive Council. The Secretary will file the annual Not-For Profit Report with the South Dakota Secretary of State.

The Secretary will have signature authority on the SDAS bank account and as necessary write checks for such things as the annual Not-For Profit Report. Checks written by the SDAS Secretary are contained within the SDAS Treasure's Report and are audited during the annual audit.

The Secretary shall provide updates to the Executive Council about the National Association of Academies of Science.

SECTION 2. The Treasurer shall collect dues, receive and disburse all funds of the Academy and have custody, along with at least one other officer, of those funds, except those described in ARTICLE X, SECTIONS 1, 4 and 5. The Treasurer is authorized to expend such funds as are necessary to conduct the business of the office. All other disbursements are to be authorized by a majority vote of the Executive Council or by a majority vote of the members present at the annual business meeting. The Treasurer shall submit the books and receipts to the members of the Auditing Committee for an annual audit.

SECTION 3. The Treasurer shall be empowered to accept such grants, gifts, and bequests as may be offered to the Academy. Stipulations as to the use of such sums as may be made by the donors shall be subject to the approval of the Executive Council.

SECTION 4. The Editor shall annually publish and distribute copies of the *Proceedings of the South Dakota Academy of Science* to all members, college libraries in South Dakota, and appropriate abstracting and indexing services. The *Proceedings* shall include the minutes of the annual meeting, a list of the current membership, obituaries of members, the Presidential Address, abstracts of papers and posters presented at the annual meeting, and full papers. He or she shall keep and update a master list of contact information of past and present members to be used to announce the call for papers; shall prepare the call for papers in conjunction with the Executive Council; shall accept submissions of abstracts of papers and posters to be presented at the annual meeting; shall prepare, print, and make available a Schedule of Events for the annual meeting; shall accept and edit abstracts and full papers; shall enlist assistant editors to help make policy decisions and establish submission and publication guidelines; shall engage experts to serve as reviewers of full papers; shall set and collect page charges; shall arrange for the formatting, printing of the Proceedings and its uploading to the Academy website; shall pay all bills related to the publication of the Proceedings and the Schedule of Events for the annual meeting; and shall maintain a record of financial transactions involving revenues and expenses.

SECTION 5. The Editor shall have charge of the distribution, sale, and exchange of the published transactions of the Academy, under such restrictions as may be imposed by the Executive Council.

ARTICLE XI
Presentation of Papers

SECTION 1. The Editor shall notify all members in ample time before any meeting at which papers may be presented, and papers to be presented must be filed with the Editor by title and abstract at a time to be designated by the Editor. Papers presented by associate members must, in addition, be co-authored, sponsored, or approved by a member. It shall be the duty of the Editor to arrange the program for the annual meeting and to provide copies of the program for the use of the members.

ARTICLE XII
Additions, Amendments, or Changes in the Articles of Incorporation and the By-Laws

SECTION 1. The Articles of Incorporation (see *Proceedings* 1989 68:142) may be amended by a three-fourths vote of members voting at an Annual Meeting, provided that notice of the desired change has been sent by the Secretary to all members at least twenty days before such meeting.

SECTION 2. By-Laws may be adopted, suspended, or amended by a three-fourths vote of the members present at any Annual Meeting of the Academy.

SECTION 3. Any changes in the Constitution (see *Proceedings* 1959 38:18) or the By-Laws of the Academy, or additions or amendments thereto, shall be made in compliance with such laws of the State of South Dakota as govern or regulate educational and scientific corporations.
SECTION 4. The rules contained in Robert's Rules of Order shall govern the Academy in all cases in which they are not inconsistent with the By-Laws of the South Dakota Academy of Science.

SDAS ANNUAL MEETING ROTATION SCHEDULE 2020-2060

Year	Location	Sponsoring Institution
2020	Cyberspace	Virtual Business Meeting, SDAS Executive Council
2021	Spearfish, SD	Black Hills State University
2022	Madison, SD	Dakota State University
2023	Vermillion, SD	University of South Dakota
2024	Aberdeen, SD	Northern State University
2025	Rapid City, SD	South Dakota School of Mines and Technology
2026	Sioux Falls, SD	University of Sioux Falls
2027	Mitchell, SD	Dakota Wesleyan University
2028	Brookings, SD	South Dakota State University
2029	Yankton, SD	Mount Marty University
2030	open	open
2031	Spearfish, SD	Black Hills State University
2032	Madison, SD	Dakota State University
2033	Vermillion, SD	University of South Dakota
2034	Aberdeen, SD	Northern State University
2035	Rapid City, SD	South Dakota School of Mines and Technology
2036	Sioux Falls, SD	University of Sioux Falls
2037	Mitchell, SD	Dakota Wesleyan University
2038	Brookings, SD	South Dakota State University
2039	Yankton, SD	Mount Marty University
2040	open	open
2041	Spearfish, SD	Black Hills State University
2042	Madison, SD	Dakota State University
2043	Vermillion, SD	University of South Dakota
2044	Aberdeen, SD	Northern State University
2045	Rapid City, SD	South Dakota School of Mines and Technology
2046	Sioux Falls, SD	University of Sioux Falls
2047	Mitchell, SD	Dakota Wesleyan University
2048	Brookings, SD	South Dakota State University
2049	Yankton, SD	Mount Marty University
2050	open	open
2051	Spearfish, SD	Black Hills State University
2052	Madison, SD	Dakota State University
2053	Vermillion, SD	University of South Dakota
2054	Aberdeen, SD	Northern State University
2055	Rapid City, SD	South Dakota School of Mines and Technology
2056	Sioux Falls, SD	University of Sioux Falls
2057	Mitchell, SD	Dakota Wesleyan University
2058	Brookings, SD	South Dakota State University
2059	Yankton, SD	Mount Marty University
2060	open	open

Complete Senior Research Papers

submitted for presentation at the

109th Annual Meeting

of the

South Dakota Academy of Science

RETHINKING SECONDARY SUCCESSION: LESSONS FROM THE 1988 FIRES IN YELLOWSTONE NATIONAL PARK

John A. Long*
Department of Science & Mathematics
Northern State University
Aberdeen, SD 57401
*Corresponding Author: john.long@northern.edu

ABSTRACT

Fire is an integral part of fire-adapted ecosystems, yet it represents a major disturbance to the landscape. For vegetative communities, a major wildfire often results in losses of habitat and core area, increases in fragmentation, and potential decreases in species richness. The 1988 fires in Yellowstone National Park were among the largest in recorded history, burning more than one-third of the park and substantial portions of the surrounding area. This paper describes landscape changes and accompanying effects on the vegetation community and structure that resulted from the 1988 fires at a detailed level by focusing on the Fan Fire in the northwestern part of the park. Results suggest that the Fan Fire induced only minor changes in the landscape. Evidence was found for a slight increase in fragmentation and a loss of core area; however, species diversity did not suffer. In general, species present before the fire were present after the fire. Specifically, there were increases in the amount of lodgepole pine, decreases in the amount of Englemann spruce, and minor changes in whitebark pine, Douglas fir, and non-forested areas. Many of the observed post-fire vegetation responses were contrary to expectations based on conventional notions of secondary succession and supported the concept of the resilience of a fire-adapted ecosystem and the importance of biotic legacies.

Keywords

Yellowstone National Park, Fan Fire, fires of 1988, secondary succession, land cover change

INTRODUCTION

Despite a substantial body of research on secondary succession, we lack a comprehensive theoretical framework. A typical description of secondary succession suggests that an ecosystem experiences change following a major disturbance that involves colonization by opportunistic organisms ultimately replaced by other species as the ecosystem approaches a steady-state, or climax, condition (e.g., Del Castillo and Blanco-Macías 2007). Pioneer species, often fast-growing and able to tolerate harsh conditions, recolonize the area first. In addition, post-disturbance change results in a more fragmented landscape in terms of land cover types and a loss of core habitat. This study offers evidence that this notion of secondary succession is not universally valid.

Fire is an integral part of some ecosystems. Although the aftermath of a major wildfire appears destructive, fire plays a vital role in ecosystem health. These ecosystems have evolved with fire as a natural component that facilitates the maintenance of an ecosystem's characteristic structure and composition. One of the more familiar fire-dependent ecosystems is the coniferous forests and associated communities of the Rocky Mountains, such as those located in the Greater Yellowstone Ecosystem (GYE) (Figure 1). Fire has long been a dominant and recurring factor in shaping vegetation distribution patterns (Knight & Wallace 1989) and recent estimates suggest that large fires recur in the GYE every 175 to 350 years (Baker 2009).

Yellowstone National Park (YNP) (Figure 2) experienced one the worst fire seasons in recorded history during the summer of 1988, eventually forcing the National Park Service to close YNP to the public for the first time since the park's establishment. By summer's end, 248 distinct fires had started in the GYE, including 16 major fires that burned approximately 3200 km^2 of YNP's nearly

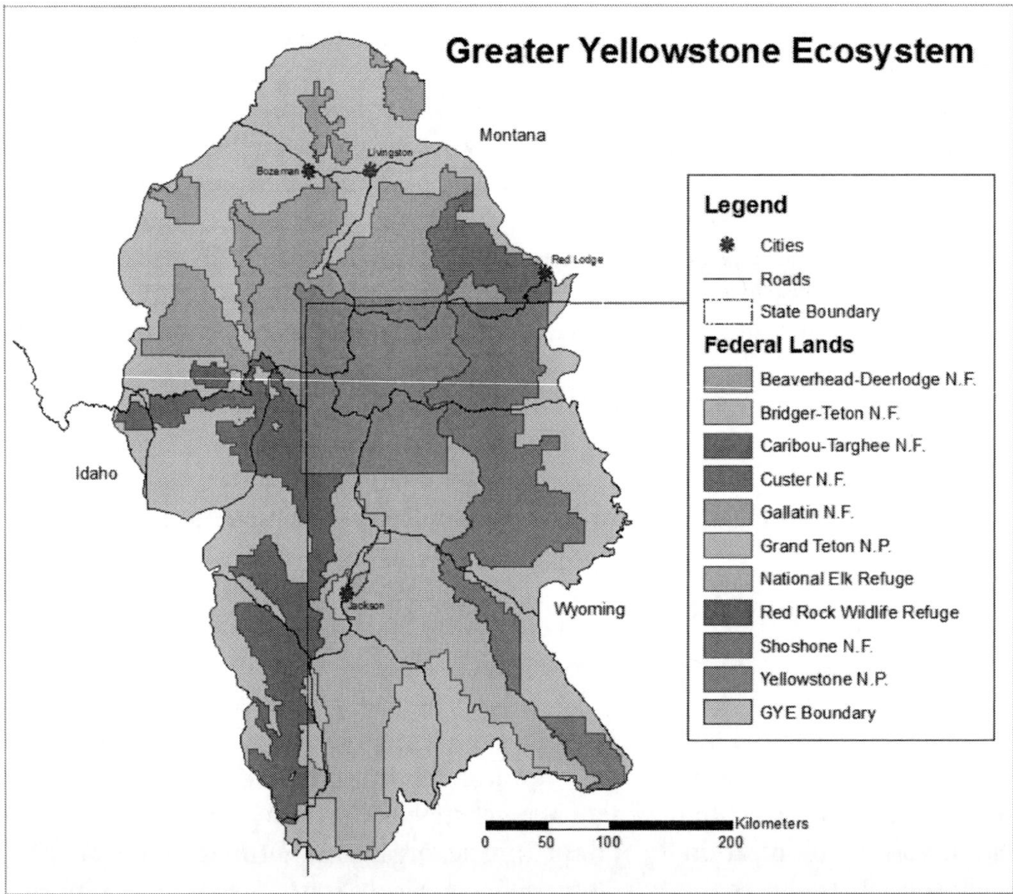

Figure 1. Map of the Greater Yellowstone Ecosystem (GYE) and associated Federal lands.

9000 km^2 (National Park Service 2013). These fires generated substantial national attention and involved the largest fire suppression effort ever undertaken to date in the United States.

Following a wildfire, an ecosystem recovers via secondary succession as the pre-fire vegetative community structure and composition experience some degree of change. This paper examined wildfire-induced changes in the landscape and the accompanying effects on vegetation community structure resulting from the 1988 fires in the GYE. The scope is limited to a single representative event – the Fan Fire, which was one of the earliest fires, igniting on June 26, and not fully extinguished until August 23. By then, it had burned 11,069 ha in the north-west corner of the park. The Fan Fire was a substantial fire, but it paled in comparison to other major fires such as the North Fork Fire that burned 214,969 ha (Rothermel et al. 1994). Specifically, the goals of this paper were to: (1) identify any changes between pre-fire and post-fire land cover types; (2) quantify any changes in terms of landscape metrics; and (3) examine the results with respect to the expected vegetative responses to major disturbance according to contemporary notions of secondary succession.

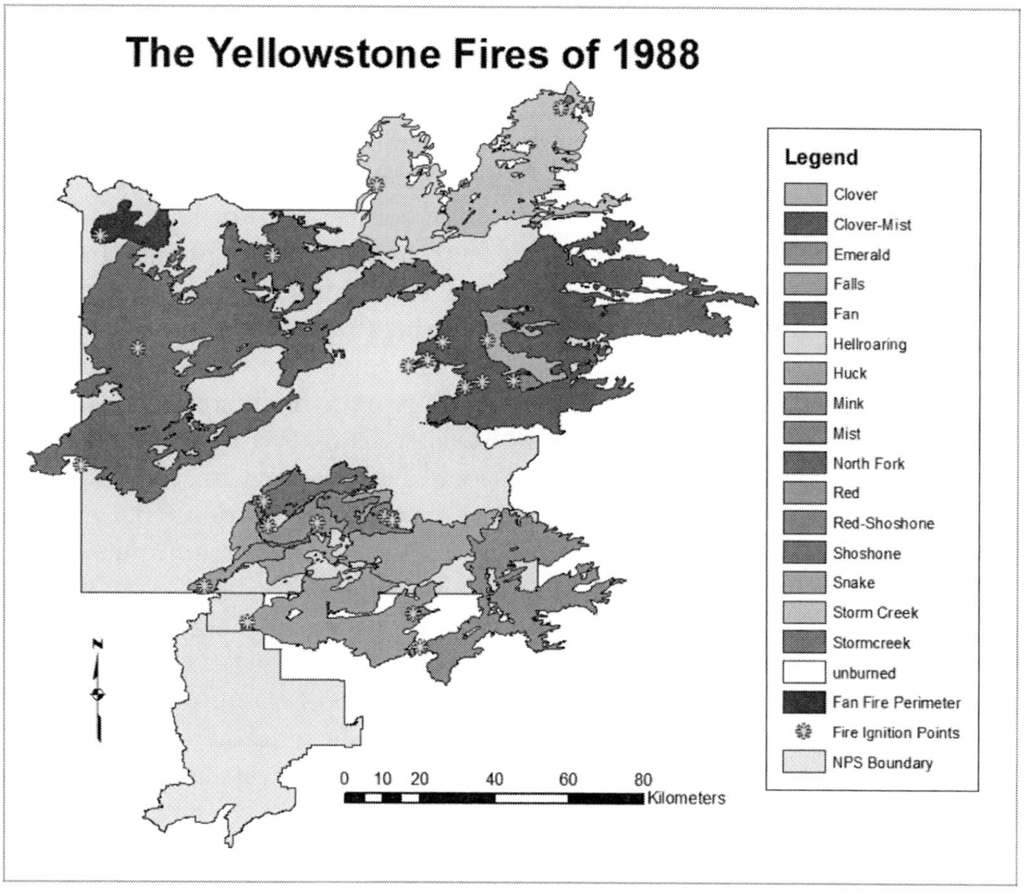

Figure 2. Locations of the major fires and ignition points in Yellowstone National Park during the 1988 fire season.

METHODS

Data were processed and analyzed with a Geographic Information System (GIS) within the ArcGIS environment (ESRI 2024). GIS data were obtained from the YNP Spatial Analysis Center and included: national park boundaries, fire start points, fire perimeters, burn types, pre-fire cover type and post-fire cover type. In addition, a 1 arc-second (30 m) digital elevation model (DEM) was obtained from the United States Geological Survey's National Elevation Dataset. The fire start points layer contained the ignition points for the major fires, accurate to within 100 m and were determined by the observations of fire lookouts. The fire perimeter layer depicted the maximum areal extent for the major fires and was determined by integrating several sources including daily infrared photography flights, satellite imagery, and ground-based reconnaissance.

The burn type layer categorized the character of the burned areas as canopy burn, mixed burn, non-forested burn, undifferentiated burn, and unburned with a 5-acre resolution (2 ha). A canopy burn denoted areas in which the forest overstory and ground cover were burned; a mixed burn described areas characterized by a combination of unburned, scorched, or blackened trees as well as burned ground cover; a non-forested burn described an area of burned ground cover only; and undifferentiated burns are burned areas that could not be reliably placed in one of the other categories.

The pre-fire cover type layer describes the land cover at 30 square meter resolution in terms of the dominant species extant prior to the 1988 fires: lodgepole pine (*Pinus contorta* var. *latifolia*), Englemann spruce (*Picea engelmannii* Parry ex Engelm.), whitebark pine (*Pinus albicaulis* Engelm.), Douglas fir (Pinus *menziesii* var. *glauca* (Beissn.) Franco), aspen, (*Populus tremuloides* Michx.), krumholtz (a stunted growth form of whitebark pine), mixed conifer, non-forested (grassland), and water. These land cover types were derived from aerial photographs and ground transects obtained between 1980 and 1988 during the summer months. The post-fire layer was created in 1991, also during the summer months, by modifying the pre-fire layer to reflect changes in land cover type following the fires.

Mapping locations with post-fire land cover types that differed from the pre-fire land cover types was a multistep process. First, land cover types were assigned numerical codes so that their categories could be converted to numerical ones to facilitate map algebra operations. Then a map of the locations that changed land cover types (Figure 3) was created by using the non-zero outputs from a spatial model in ArcGIS in which the code for the pre-fire cover type was subtracted from the code for the post-fire cover type using map algebra operations in the GIS system. When the pre-fire and post-fire cover types were the same, the output from the spatial model was zero, while non-zero outputs indicated a change in land cover type. This process identified the changed areas but was not capable of identifying the specific transition because the non-zero outputs were not unique. For example, land cover type 5 minus land cover type code 2 and land cover type 4 minus land cover type code 1 each yielded a region coded as 3, but they clearly represented different transitions.

Next, a more explicit map (Figure 4) was made of the changed areas in terms of specific transitions (e.g., Douglas fir to lodgepole pine). The process was the same as with the general change map except that the post-fire cover type codes were multiplied by ten; thus, upon subtraction, each possible transition had a unique code. The unique land cover type codes allowed the mapping of specific changes.

Landscape metrics are important tools for measuring and describing spatio-temporal patterns and structures, making them ideally suited to examine changes in vegetation structure and composition. Since landscape metrics can vary by scale, they were computed for the region in general, hereafter referred to as the landscape level, for both pre-fire and post-fire conditions, as well as for specific species, hereafter referred to as class level. All landscape measures were computed using the landscape analysis tool V-LATE 2.0, an ArcGIS extension.

Metrics associated with spatial configuration include those related to patch size and shape, the connectivity of the patches, and their locations with respect to neighboring patches. Metrics associated with landscape composition include those related to area, richness, and diversity. Specifically, this study measured: total area (TA), class area (CA), class area as a proportion of the total area (CP), number of patches (NP), mean patch size (MPS), total core area (TCA), number of core areas (NCA), total class core area (TCCA), core area index (CAI), total edge (TE), mean patch edge (MPE), mean shape index (MSI), mean perimeter-area ratio (MPAR), species richness (S), Shannon's diversity index (SDI), Shannon's evenness index (SEI), and the dominance index (D_1). These measures were calculated at the landscape level, class level or both as appropriate. Definitions, methods of calculations and importance of these measures can be found in many references (e.g., Leitão et al. 2006) for readers unfamiliar with landscape metrics.

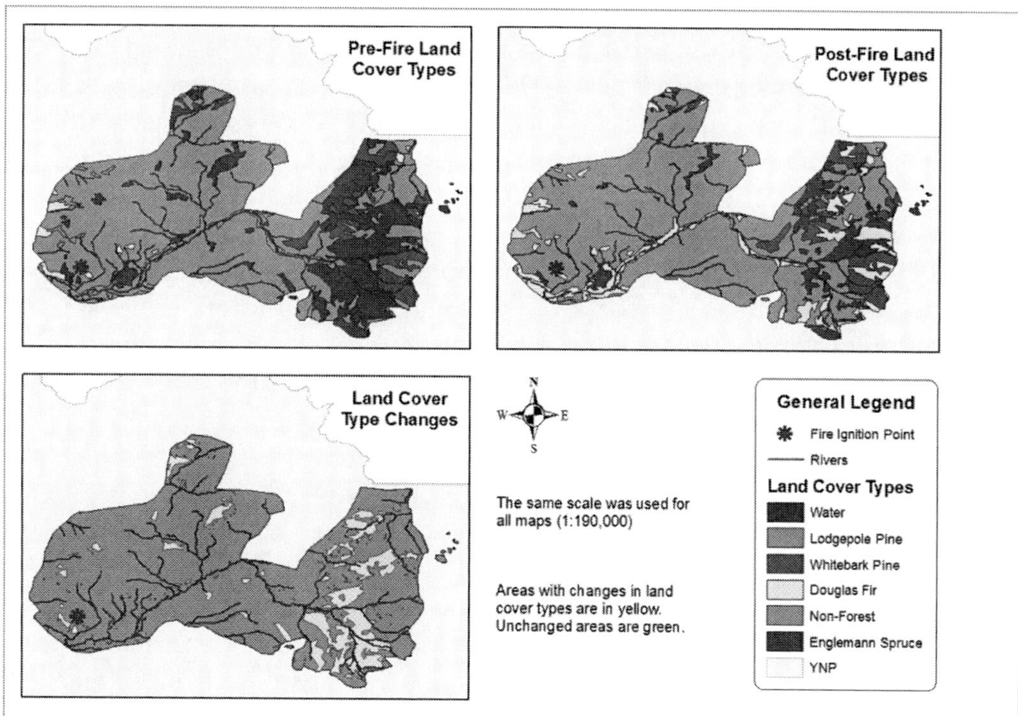

Figure 3. Map of the pre-fire and post-fire cover types, and land cover changes within the Fan Fire perimeter.

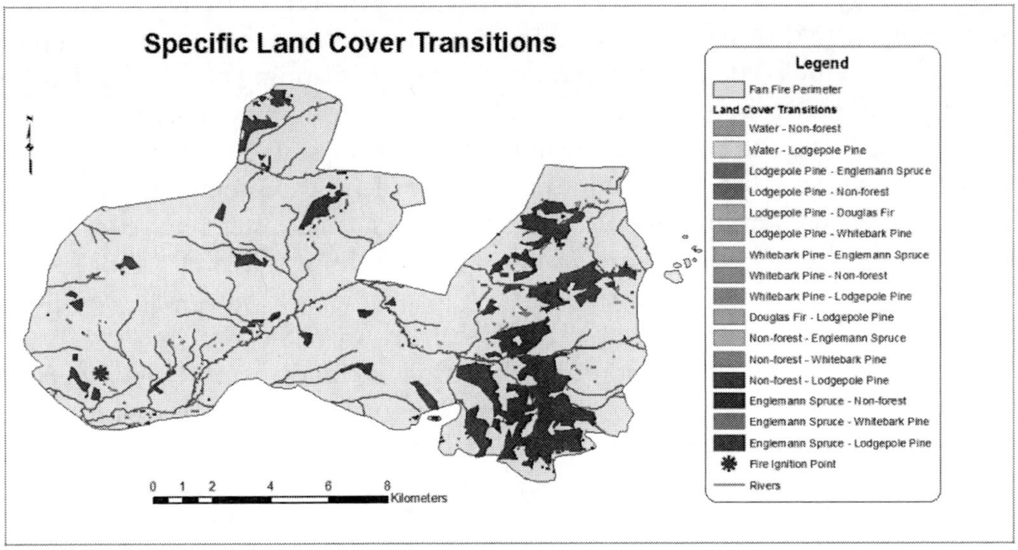

Figure 4. Map of specific land cover type changes within the perimeter of the Fan Fire.

RESULTS

Yellowstone National Park is located near the center of the GYE, and the Fan Fire burned the extreme northwest corner of the park. The pre-fire vegetation structure in this region of YNP was dominated by vast areas of lodgepole pine, interspersed with whitebark pine and Englemann spruce at higher elevations, or Douglas-fir at lower elevations. Small areas of aspen and sizeable areas of grassland were also present in the area. This vegetative composition was the case within the perimeter of the Fan Fire except that there were no areas of aspen.

The Fan Fire encompassed nearly 11,070 ha, and approximately 1338 ha (12.1%) of that area experienced a change in land cover type. Nearly 88% of the area experienced no change in land cover type despite experiencing a major disturbance. The post-disturbance colonization by pioneer species as suggested by many did not materialize. Pre-fire land cover types within the perimeter of the Fan Fire consisted of six of the nine possible YNP land cover types: lodgepole pine, Englemann Spruce, whitebark pine, Douglas fir, non-forest, and water. Aspen, krummholz, and mixed conifer were not present within the confines of the Fan Fire perimeter. The post-fire land cover types were the same as for the pre-fire land cover types, albeit in different proportions and locations.

Specific land cover changes are represented as a change matrix (Table 1) in terms of the number of pixels (30 m^2 = 0.003 ha per pixel) experiencing change and as a percentage of the changed area (Table 2). Sixteen of the 30 possible land cover transitions occurred; however, more than 95% of the area that changed land cover type was a result of three specific transitions. The overwhelming majority (91.93%) of land cover changes were in areas that were Englemann spruce pre-fire that transitioned to lodgepole pine post-fire. Transitions from Englemann spruce to non-forested accounted for 2.04% of changes and non-forested to lodgepole pine accounted for 1.06%. Since the data were derived from aerial imagery, they represented a census of the area and were not sample-based; therefore, observed differences are considered true differences and any discussion of statistical significance is inappropriate. The remaining 13 transitions accounted for less than one percent each and were likely artifacts of the raster to polygon conversion process as they were typically single pixels located along the edges of a land cover type. Most of the land cover changes took place in the eastern portion of the Fan Fire perimeter, where the majority of the Englemann spruce was prior to the fire.

There was evidence for small post-fire changes at the landscape level including an increase in fragmentation, loss in core area, and a loss of diversity. The increase in fragmentation was supported by several metrics including: (1) an increase in the number of patches (NP) from 103 to 113; (2) an increase in total edge (TE); (3) a decrease in mean patch edge (MPE); (4) a slight increase in the mean shape index (MSI); and (5) a decrease in the mean perimeter-area ratio (MPAR). The loss of core area was minor and was supported by a 16-ha decrease in total core area (TCA) and a corresponding slight reduction in core area index

Table 1. Change matrix in terms of the number of pixels changing land cover type. Land cover types: W = water, LP = lodgepole pine, WB = whitebark pine, DF = Douglas fir, NF = non-forested, ES = Englemann Spruce. Total pixels = 57,212. Total unchanged pixels = 50,297. Total changed pixels = 6915. Pixels can be converted to area by multiplying by 30 m^2 per pixel or to hectares by multiplying by 0.003 ha per pixel. Values on the diagonals represent unchanged land cover types.

		To					
		W	LP	WB	DF	NF	ES
From	W	10	4	0	0	9	0
	LP	0	35705	14	8	65	43
	WB	0	42	3605	0	34	34
	DF	0	29	0	642	0	0
	NF	0	73	8	0	5371	14
	ES	0	6357	40	0	141	4964

Table 2. Land cover type changes as a percent of the total number of changed pixels. Land cover types: W = water, LP = lodgepole pine, WB = whitebark pine, DF = Douglas fir, NF = non-forested, ES = Englemann Spruce, - = no change in land cover type.

		To					
		W	LP	WB	DF	NF	ES
From	W	-	0.06	-	-	0.13	-
	LP	-	-	0.20	0.12	0.94	0.62
	WB	-	0.61	-	-	0.49	0.49
	DF	-	0.42	-	-	-	-
	NF	-	1.06	0.12	-	-	0.20
	ES	-	91.93	0.58	-	2.04	-

(CAI). Although species richness (S) did not change, there was evidence for a loss of diversity from: (1) a decrease in Shannon's Diversity Index; (2) a decrease in Shannon's Evenness Index; and (3) and an increase in dominance. These results are summarized in Table 3. Given that there were only six landcover types in the affected region, these changes reflect changes in the proportions of the extant species rather than a change in species composition following the Fan Fire.

Analysis at the class-level (individual species) for lodgepole pine indicated that the class proportion had increased from 62.6% to 73.7% of the area within the Fan Fire perimeter. This was accompanied by an increase in the number of core areas (NCA) as well as a 17.7% increase in core area (CA). On the other hand, lodgepole pine showed a slight increase in fragmentation as evidenced by increases in NP, NCA, and TE, as well as decreases in mean patch size (MPS) and MPE. Practically all the expansion in lodgepole pine came at the expense of Englemann spruce as the overwhelming majority (97.2%) of the area lost by Englemann spruce transitioned to lodgepole pine after the fire. The class proportion for Englemann spruce dropped from 20.1% of the area pre-fire to 8.9% in the post-fire landscape. In addition, Englemann spruce lost more than half of its core area and what remained became substantially more fragmented.

Similar lines of evidence from the class-level metrics for the other species produced the following results. Whitebark pine experienced a small change (0.1%) in class proportion, with a slight loss in core area (10 ha) that was accompanied by a *decrease* in fragmentation. This is the opposite pattern to that of lodgepole

Table 3. Landscape level metrics. All areas are reported in hectares and lengths are in meters. Indices are unitless. Metrics: TA = total area, NP = number of patches, TCA = total core area, CAI = core area index, S = species richness, SDI = Shannon's Diversity Index, SEI = Shannon's evenness index, D_I = dominance index, TE = total edge, MPE = mean patch edge, MSI = mean shape index, and MPAR = mean perimeter-area ratio.

	Metric	Pre-Fire	Post-Fire
Area	TA	11,068 ha	11,068 ha
	NP	103	113
Core Area	TCA	10,576 ha	10,560 ha
	CAI	95.55%	95.41%
Diversity	S	6	6
	SDI	1.073	0.897
	SEI	0.599	0.501
	D_I	0.719	0.895
Edge	TE	499,955 m	516,805 m
	MPE	4854 m	4573 m
Form	MSI	1.666	1.699
	MPAR	0.026	0.022

pine. Non-forested areas increased slightly (0.25%) in class proportion and gained 2.6% in core area; however, this class also experienced large decreases in fragmentation suggested by increases in core area. Areas of Douglas fir and water encountered essentially no change as measured by any of the class-metrics. Details for class-level metrics are summarized in Table 4.

Table 4. Class level metrics. All areas are reported in hectares and lengths are in meters. Indices are unitless. Metrics: NP = number of patches, MPS = mean patch size, NCA = number of core areas, CA = core area, TCCA = total class core area, CAI = core area index, CP = class area as a proportion of the total area, TE = total edge, MPE = mean patch edge, MSI = mean shape index, and MPAR = mean perimeter-area ratio.

		Lodgepole Pine		Englemann Spruce		Whitebark Pine	
	Metric	Pre-fire	Post-fire	Pre-fire	Post-fire	Pre-fire	Post-fire
Area	NP	8	11	27	45	16	15
	MPS	867 ha	742 ha	82 ha	22 ha	45 ha	47 ha
Core Area	NCA	17	19	41	55	18	17
	CA	6933 ha	8158 ha	2221 ha	983 ha	721 ha	711 ha
	TCCA	6761 ha	7948 ha	2095 ha	874 ha	666 ha	657 ha
	CAI	97.52%	97.43%	94.32%	88.88%	92.36%	92.43%
Diversity	CP	62.64%	73.7%	20.07%	8.88%	6.51%	6.42%
Edge	TE	172,117 m	210,667 m	128,802 m	112,912 m	55,975 m	54,745 m
	MPE	21,515 m	19,152 m	4770 m	2509 m	3498 m	3650 m
Form	MSI	2.209	2.031	1.623	1.668	1.637	1.641
	MPAR	0.021	0.021	0.032	0.029	0.018	0.012

		Douglas Fir		Non-forest		Water	
		Pre-fire	Post-fire	Pre-fire	Post-fire	Pre-fire	Post-fire
Area	NP	10	10	40	31	2	1
	MPS	12.8 ha	12.6 ha	25.5 ha	35.1 ha	2.2 ha	2.2 ha
Core Area	NCA	10	10	51	42	2	1
	CA	128 ha	126 ha.82	1060 ha	1088 ha	4.4 ha	2.2 ha
	TCCA	113 ha	111 ha	938 ha	969 ha	3.3 ha	1.6 ha
	CAI	88.42%	88.35%	88.41%	89.01%	74.65%	73.04%
Diversity	CP	1.16%	1.14%	9.58%	9.83%	0.04%	0.02%
Edge	TE	15,278 m	15,109 m	126,590 m	122,720 m	1194 m	652 m
	MPE	1528 m	1511 m	3165 m	3959 m	597 m	652 m
Form	MSI	1.259	1.257	1.726	1.812	1.136	1.228
	MPAR	0.015	0.015	0.029	0.02	0.027	0.029

DISCUSSION

The Fan Fire did lead to changes in the pre-fire vegetative structure; however, the extent of the changes was minimal, and species diversity was unaffected. The overall effect was a marginally more fragmented post-fire landscape with an accompanying reduction in core area and increase in the proportion of the landscape with lodgepole pine. In total, the results of this study simultaneously support and refute the conventional notion of how secondary succession proceeds following a major wildfire.

The finding that more than 95% of the area that changed land cover type resulted from only three distinct transitions makes sense in the context of the fire-adaptability of the species. Ecological succession following a major disturbance can proceed along different trajectories. The initial stages of secondary succession are influenced by the recruitment and immigration of mobile organisms, as well as the persistence and growth of surviving organisms (Dale et al. 2005). Indeed, the observed post-fire response was initially led by the biotic legacies – such as seeds – and regenerative structures such as deep roots of the pre-fire species.

The forests of the GYE are uniquely adapted for post-fire survival. Lodgepole pine is a thin-barked species that easily burns, but deep tap roots in mature trees provide some degree of insulation from the intense heat (Anderson 2003). Some trees employ serotinous cones – cones glued shut with a strong resin. Seeds are protected within the cone and fire is required to melt the resin to release the seeds. Serotiny is a seed-dispersal strategy that favors survivorship since it allows for maximum seed dissemination immediately after a fire, when resource availability is high and inter-species competition is low (Aber & Melillo 2001).

Serotiny is not ubiquitous in lodgepole pine (e.g., Knight & Wallace 1989). Whether or not a tree is serotinous depends on elevation and the age of the tree (Schoennagel et al. 2003). In the GYE, serotiny is uncommon for trees above 2400 m and for young trees below 2400 m; thus, serotiny is most prevalent in older lodgepole pines below 2400 m (Schoennagel et al. 2003). The region within the Fan Fire ranges from 1350 to 4200 m in elevation with a nearly even split between elevations below and above 2400 m. Serotinous and non-serotinous trees often coexist and their respective proportions determined by the nature and timing of the last wildfire (Knight & Wallace 1989). Since fire suppression was park policy in Yellowstone for the majority of the twentieth century (e.g., Zimmerman 2009), it is not surprising that serotiny is less common in younger trees (< 75 years).

In contrast to lodgepole pine, Englemann spruce is a fire-sensitive species and generally does not survive even low-intensity fires (Uchytil 1991). Englemann spruce has a thin layer of bark, offering little protection for the cambium. Shallow roots offer little protection from ground fire while the flammable foliage is highly susceptible to crown fire. Post-fire reestablishment of Englemann spruce relies on wind-dispersed seeds from surviving stands (Uchytil 1991). The fire suppression legacy in YNP allowed Englemann spruce to produce dense stands vulnerable to wildfire.

Lodgepole pine has been very successful at reestablishing areas that were primarily lodgepole pine prior to the fire, as well as colonizing substantial areas that were previously Englemann spruce. The overwhelming majority (91.93%) of transitions in land cover type were Englemann spruce to lodgepole pine, while Englemann spruce to non-forested areas accounted for 2.04% and non-forested to lodgepole pine accounted for 1.06%. Thus, the fundamental observation here is the superiority of lodgepole pine in secondary succession versus the inability of Englemann spruce to compete; both are direct results of the fire-adaptive nature of lodgepole pine and the fire-sensitivity of Englemann spruce. Since much of the area that did experience a land cover type change involved the loss of Englemann spruce, this suggests that recolonization from unaffected regions played a very small role in the post-fire recovery.

The relatively small changes in area-based metrics in comparison to lodgepole pine for the other classes reflect species-specific adaptation strategies to increase post-fire survivorship. Whitebark pine, a high-elevation specialist, and a vital food resource for GYE grizzly bears, uses a soil seed-bank strategy in which seeds are cached underground by foraging birds (Fryer 2002), especially Clark's nutcracker (*Nucifraga columbiana* (Wilson)). Depending on the depth of burial, seeds can often survive an intense fire. Douglas fir typically grows at lower elevations where the fires were less intense and mature trees were aided by a thick insulating bark that increased survival (Steinberg 2002). In addition, most of the shrubs, forbs, and graminoids present before the fire resprouted readily (Baker 2009) and were seen as early as the following spring (Turner 2009). Since the above ground structures of shrubs forbs, and grasses typically do not survive even low intensity fire (Morgan & Neuenschwander 1988), survival of these non-tree plants depends on their below ground structures including: buds, roots, rhizomes, bulbs etc. Again, the recolonization from unaffected regions played a minimal role in the post-fire recovery.

The GYE is not notably diverse in the number of extant species and the species richness was unaffected by the fire. Changes in vegetation structure suggest that pre-fire land cover type and biomass contribute substantially to the regrowth potential of vegetation. For example, locations with high concentrations of lodgepole pine pre-fire generally had high post-fire concentrations because: (1) the environment continues to favor the species; (2) locations with high pre-fire biomass left more biotic legacies (e.g., seeds from serotinous cones, unburned root systems); and (3) those biotic legacies are adapted to recolonize a burned area. Similar cases can be made for Douglas fir, whitebark pine, and the non-forested areas. Burn severity, or fire intensity (Figure 5), had very little, if any, effect on regrowth across land cover types, which is not surprising as the GYE is a fire-adapted ecosystem.

Although ecological interactions are complex, some post-fire responses were expected. For example, lodgepole pine reestablishment was ample and the sapling density was the greatest at lower elevations, consistent with trends in serotiny (Turner 2009). On the other hand, post-fire recovery produced some unexpected outcomes. For example, native plant cover was rapidly restored by natural processes, which was contrary to several notions. First, some predicted that recolonization would be slow, especially for tree species, with estimates ranging from a few years for lodgepole pine to many decades for other species (see e.g., Knight & Wallace 1989 or Turner 2009 for further discussion). As discussed earlier, contemporary theory also suggested that recolonization would be dependent on long distance dispersal from unburned areas (Turner 2009), which turned out to not be the case. It was also expected that non-native plants, especially herbaceous vegetation, would invade the burned areas, but in fact they managed to establish themselves in less than 3% of the burned areas in the GYE, typically along roadways (Turner 2009). Finally, there were some assumptions that such a devastating disturbance would have a homogenizing effect on the landscape (Baskin 1999). This assumption turned out to be incorrect as suggested by the landscape metrics which indicated a slightly more fragmented landscape, but one that retained all its species. Homogenization simply did not occur.

The major landscape alterations: increased lodgepole pine, decreased Englemann spruce, and relatively minor changes in whitebark pine, Douglas fir, and non-forested areas, support the concept of the resilience of a fire-adapted ecosystem and the importance of their biotic legacies. With the single exception of the inverse relationship between amounts of lodgepole pine and Englemann spruce, pre-fire and post-fire land cover types showed very little change; species that were present pre-fire were the ones that recolonized the area post-fire. In this study, the pre-fire and post-fire data are three years apart. Subsequent studies examining changes in land cover type over a longer time suggest that the conclusions are robust. Turner (2016) reports that 24 years after the 1988 fires, lodgepole pine continues to dominate the regrowing forest.

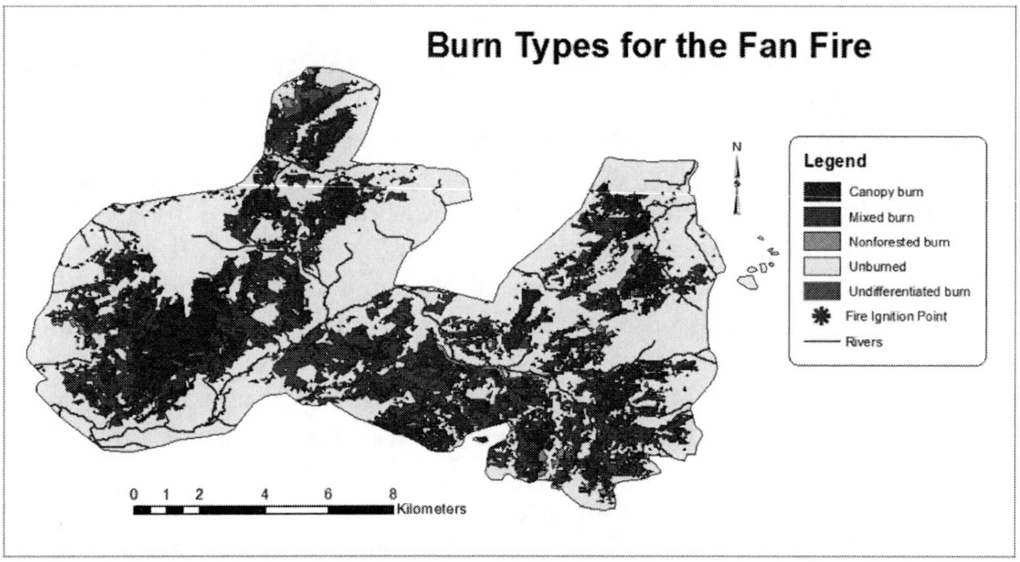

Figure 5. Map of burn types within the perimeter of the Fan Fire.

LITERATURE CITED

Aber, J. and J. Melillo. 2001. Terrestrial ecosystems (2nd ed.). Academic Press, San Diego, CA.

Anderson, Michelle D. 2003. Fire Effects Information System [Online], Pinus contorta var. latifolia. https://www.fs.usda.gov /database/feis/plants/tree/pinconl/all.html [Cited 28 Feb 2024].

Baker, W. 2009. Fire ecology in Rocky Mountain landscapes. Island Press, Washington, D.C.

Baskin, Y. 1999. Yellowstone fires: a decade later. BioScience 49 (2): 93-97.

Del Castillo, R. del and A. Blanco-Macías. 2007. Secondary succession under a slash-and-burn regime in a tropical montane cloud forest: soil and vegetation characteristics. Pages 158-180 in Newton, A.C., editor. Biodiversity loss and conservation in fragmented forest landscapes. The forests of montane Mexico and temperate South America. CABI, Wallingford, Oxfordshire, UK.

Dale, V., F. Swanson, and C. Crisafulli. 2005. Disturbance, survival, and succession: understanding ecological responses to the 1980 eruption of Mount St. Helens. Pages 3-12 *in* Dale, V., F. Swanson, and C. Crisafulli, editors. Ecological response to the 1980 eruption of Mount St. Helens. Springer, New York.

ESRI, 2024. ArcGIS Pro 3.2. Environmental Systems Research Institute (ESRI), Redlands, CA.

Fryer, Janet L. 2002. Fire Effects Information System [Online], Pinus albicaulis. https://www.fs.usda.gov/database/feis/plants/tree/pinalb/all.html [Cited 28 Feb 2024].

Knight, D., and L. Wallace. 1989. The Yellowstone fires: issues in landscape ecology. BioScience 39(10): 700-706.

Leitão, A., J. Miller, J. Arhern, and K. McGarigal. 2006. Measuring landscapes: A planner's handbook. Island Press, Washington, D.C.

Morgan, P., and L.F. Neuenschwander. 1988. Shrub response to high and low severity burns following clearcutting in northern Idaho. Western Journal of Applied Forestry 3: 5–9.

National Park Service. 2013. Yellowstone National Park, 1988: a 25th anniversary retrospective. https://www.nps.gov/articles/wildland-fire-yell-1988-25th-anniv-retrospective.htm [Cited 20 Feb 2024].

Rothermel, R., R. Hartford, and C. Chase. 1994. Fire growth maps for the 1988 Greater Yellowstone Area fires. Gen. Tech. Rep. INT-204. US Department of Agriculture, Forest Service, Intermountain Research Station, Odgen, UT.

Schoennagel, T., M. Turner, and W. Romme. 2003. The influence of fire interval and serotiny on postfire lodgepole pine density in Yellowstone National Park. Ecology 84(11): 2967-2978.

Steinberg, Peter D. 2002. Fire Effects Information System [Online], Pseudotsuga menziesii var. glauca. https://www.fs.usda.gov /database/feis/plants/tree/psemeng/all.html [Cited 26 Feb 2024].

Turner, M. 2009. Ecological effects of the '88 Yellowstone fires. Yellowstone Science 17 (2): 24-30.

Turner, M. 2016. Twenty-four years after the Yellowstone fires: Are postfire lodgepole pine stands recovering in structure and function? Ecology 97(5): 1260-1273.

Uchytil, R. J. 1991. Fire Effects Information System [Online], Picea engelmannii. https://www.fs.usda.gov/database/feis/plants/tree/piceng/all.html [Cited 27 Feb 2024].

Zimmerman, T. 2009. Wildland fire management policy. Yellowstone Science 17 (2): 31-34.

LONG- AND SHORT-TERM WEATHER VARIABLES AFFECT CAPTURE RATES IN WESTERN PAINTED TURTLES (*CHRYSEMYS PICTA BELLII*) IN MARSHALL CO., SOUTH DAKOTA

Amy Dolan[1*], Heather Waye[2], and Peter Dolan[2]
[1]Department of Science and Mathematics
Northern State University
Aberdeen, SD 57401
[2]Division of Science and Mathematics
University of Minnesota Morris
Morris, MN 56267
* Corresponding author: amy.dolan@northern.edu

ABSTRACT

Turtle populations are declining globally, and yet accurate long-term population surveys are often lacking. Even amongst widespread species, information on long-term population trends is largely missing. As such, studies on demographics, population structure, and factors that affect capture efficacy are needed. In 2019 we began a long-term study on Western Painted Turtles (*Chrysemys picta bellii*) in the Couteau des Prairie habitat in northeastern South Dakota. We captured turtles using several trapping techniques and investigated the effects of short- and long-term weather variables on catch per unit effort. Temperature was positively associated with daily capture rates, showing the strongest relationship with the temperature on the day of capture, and less so with the temperature on the day preceding capture. All other short-term weather conditions did not exhibit a statistically significant association with trapping efficacy. These data suggest that increased temperature may increase turtle movement, thus increasing chances of capture. Total monthly precipitation was strongly and negatively correlated with total catch per unit effort, showing the highest correlation with precipitation in the previous six months. We hypothesize that dry conditions favor turtle captures due to contraction of aquatic habitat, serving to concentrate turtles in the study area.

INTRODUCTION

Freshwater turtles are declining globally, with an estimated 60% of species threatened, endangered or extinct (Lovich et al. 2018). Life history traits such as slow growth rates, long generation times and high hatchling and juvenile mortality rates all leave turtles particularly vulnerable to anthropogenic changes to the climate and environment (Lovich et al. 2018; Stanford et al. 2020). Indeed, habitat loss and fragmentation due to urbanization have been shown to negatively affect numerous turtle species (reviewed in Stanford et al. 2020). These same life history traits make them useful as environmental indicators and have been shown to be useful for studies of environmental contamination (reviewed in dos Santos et al. 2021). Congruently, climatic variables have been shown to affect many aspects of reproductive success including turtle nesting phenology (Hederick et al. 2021), incubation duration (Santoro et al. 2023), and hatchling and adult survival (Santoro et al. 2023; Beard and Powell 2023).

Turtles are an important part of the ecosystem, contributing to energy flow, and significantly influencing the biotic and abiotic environment (Lovich et al. 2018). Researchers must determine population sizes and track them over time to assess the impact of anthropogenic and other pressures on freshwater turtles. Most commonly, mark-recapture studies are employed to estimate populations (Krebs 1999), however protocols vary widely by study. Furthermore, many factors have been shown to affect trap efficacy (reviewed in Mali et al. 2014) including trap type (Waye et al. 2023; Gamble 2006; Tesche and Hodges 2015), timespan of a trap in one location (Hollander et al. 2022), hoop net diameter (Ennen et al. 2021), type and tenure of bait (Mali et al. 2012) and trap mesh size (Ennen et al. 2021). These, and other, trapping biases may influence population estimates and misrepresent population size or demographic characteristics (Tesche and Hodges 2015). Thus, accurate population surveys require knowledge of how both study design and external factors influence capture probabilities.

Western Painted Turtles (*Chrysemys picta bellii* Gray 1830) are small, semi-aquatic, habitat generalists that reside in lakes, ponds and slow-moving streams and require access to loose sandy substrates for nesting. Turtles in northern populations spend winter months under the ice, which is possible due to their low metabolic rate in cold water and specific adaptations for hibernation in low-oxygen environments (Jackson 2002). When the ice melts, they leave the hibernation sites and move to other water bodies. Turtles gradually move back to the hibernation pond over the summer, and into deeper water heading into winter (Ernst and Lovich 2009). Although this species is widespread across much of North America, it remains relatively understudied in the Coteau des Prairies habitat in NE South Dakota as current information relies on bycatch data (Moos and Blackwell 2016, 2017) and voucher specimens (Davis 2023). In 2019, we began a long-term study of Western Painted Turtles in Marshall County, SD. We collected weather data and used multiple trapping techniques to carry out a multi-year mark-recapture study. Our overall aim in this work is to track population changes over time and examine the anthropogenic influences on turtle populations. To do this requires accurate population estimates. Hence, this particular study aims to assess the impact of both long- and short-term weather variables on capture probabilities of painted turtles.

METHODS

Clear Lake is located in Marshall County, South Dakota, 27 km east of Sisseton, SD, and is surrounded by hay and pastureland. The lake is heavily used for recreation. The surrounding shoreline is occupied by 207 residences, with a trend of increased year-round occupancy and size of structures (Helms 2010). The lake is fed by two inlets, one from Long Lake to the north and one from North Red Iron Lake to the east. Highway 10 flanks the eastern edge of the lake.

We established our study site (near 45.695469, -97.340790) in 2019 on the southeastern corner of the lake where the inlet pond from North Red Iron Lake enters Clear Lake on the east side of Highway 10 (Figure 1). This site is flanked by both public and private land and surrounded by a riparian habitat of trees, bushes, grasses and many natural basking spots. In 2019 to 2022 we used basking traps that were 71 cm x 71 cm x 30 cm (#840879 Sundeck from Heinsohn's Country Store, www.texastastes.com), and hoop traps that were 91 cm diameter 3 hoop (#TN315 from Memphis Net & Twine), about 180 cm long with 3.8 cm mesh. In 2023 we constructed basking traps to include a middle basking platform and replaced the hoop traps with ones with the same measurements but 2.5 cm mesh.

We set multiple basking traps and hoop traps for a duration of at least one-week (Table 1). Traps were checked at approximately 24-hour intervals. Traps were deployed between 7 and 24 days depending on the year (2019: 6/21 to 7/13; 2020: 6/15 to 7/9; 2021: 6/14-6/21; 2022: 6/23-6/30 and 2023: 6/13-6/20). Occasionally, hoop traps became damaged to an extent that they were not able to be repaired in the field. In those cases, traps were removed, temporarily reducing our trapping effort.

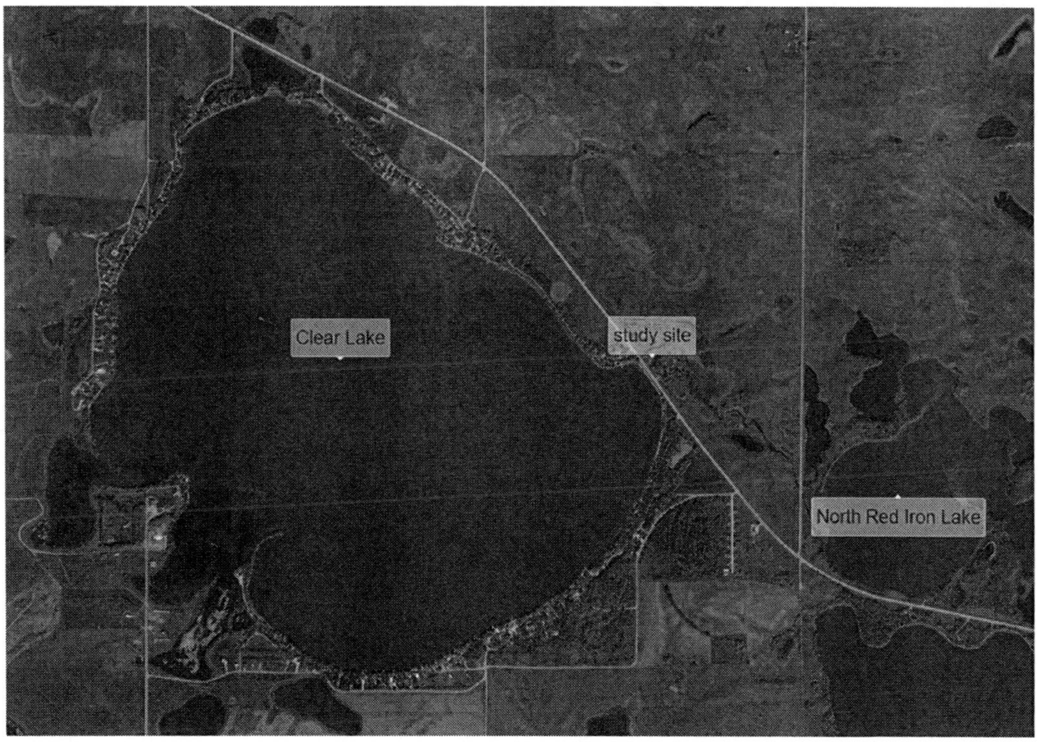

Figure 1. Aerial photo of Clear Lake, Marshall County, South Dakota, including North Red Iron Lake and the location of the study site (Waye et al. 2023).

Captured turtles were marked with a unique color-number combination using a non-toxic oil-based marker (Kornilev et al. 2012). Mass was recorded to the nearest 5 grams, and Straight Carapace Length (SCL), Body Width (BW) and Body Depth (BD) were recorded in millimeters. Sex was determined for all individuals whose SCL >50 mm by comparing the location of the cloaca relative to the edge of the plastron, the appearance of tomiodonts and the length of foreclaws. Individuals with an SCL <50 mm were recorded as juveniles. Pictures of the plastron and carapace as well as blood samples were taken for additional studies.

Weather variables were obtained from the National Weather Service from the nearest weather station in Sisseton, SD (https://www.weather.gov/wrh/climate?wfo=abr). Long-term weather data included total monthly precipitation in the previous year, previous six-months, and during the winter (December through March as per Moos and Blackwell 2016). Short-term weather data included precipitation and high, low, and average temperature on sampling days. These values were used in four ways:

1. Value on day of capture
2. Value on day preceding capture
3. Average of three days preceding capture
4. Average of seven days preceding capture.

For examination of short-term climate variables, we calculated catch per unit effort (CPUE) for basking traps, hoop traps, and the total of the two for each day (total captures per day/total traps). The statistical significance of the correlation coefficients were examined using the R function cor.test() which performs a hypothesis test with a null hypothesis of $r = 0$. Low *P*- values indicate statistical significance of the correlation coefficient, *r*. *P*-values of ≤0.05 were considered significant.

RESULTS

Total annual number of turtles captured varied between 7 and 51, and CPUE ranged from 0.055 turtles per trap-day in 2019 to 0.708 turtles per trap-day in 2020. Hoop traps were the most effective capture method both in terms of total turtles captured and CPUE; only one year, 2020, had a relatively large number of basking trap captures (Table 1).

Short-term weather: Precipitation—Almost two-thirds of the study days had no precipitation (43/64), while most of the remaining days (18/64) had only slight accumulation (defined as less than 12 mm of accumulation in a day) and only 3 days had higher rainfalls. As a result, no meaningful relationships between short-term rainfall and CPUE were detected.

Short-term weather: Temperature—Across multiple years, total daily CPUE produced more consistent linear patterns than basking CPUE or hoop CPUE alone. We presume that the same set of turtles was being exposed to both trap types, and a turtle caught by a basking trap was then not available to be captured that day by a hoop trap and vice-versa. Finally, since only 2020 had a significant number of basking captures, we used total CPUE for our linear regressions. The *P*-value in a linear regression is testing (t-test) the null hypothesis that the value of the coefficient is 0.

For each temperature variable, linear models were built to predict total CPUE. Additional explanatory indicator variables were included to capture differences in year-to-year baselines. The *P*-values were recorded for each coefficient (Table 2). The *P*-value in the row variable column captures statistical significance of the corresponding temperature variable on total CPUE. The temperature for the previous day and for the capture days show a strong statistically significant positive association with total CPUE (Figure 2).

Previous day and capture day temperature variables are strongly to weakly correlated with each other (across the fifteen variable pairs Pearson correlation varied from $r = 0.424$ to $r = 0.891$). Even in the case of the weakest correlation, the multicollinearity masks statistical significance in one or more of the temperature variables, and including such extra variables doesn't elucidate any underlying relationship between temperature and total CPUE. The underlying message in all models is the same: temperatures near time of capture are positively correlated with increased CPUE (e.g. capture day average, Figure 3)

There is one CPUE outlier from 2020 (Figure 2). With this observation, the distribution of the residuals in all models are statistically different from a normal distribution (Shapiro Wilkes test of normality). With this outlier removed, the

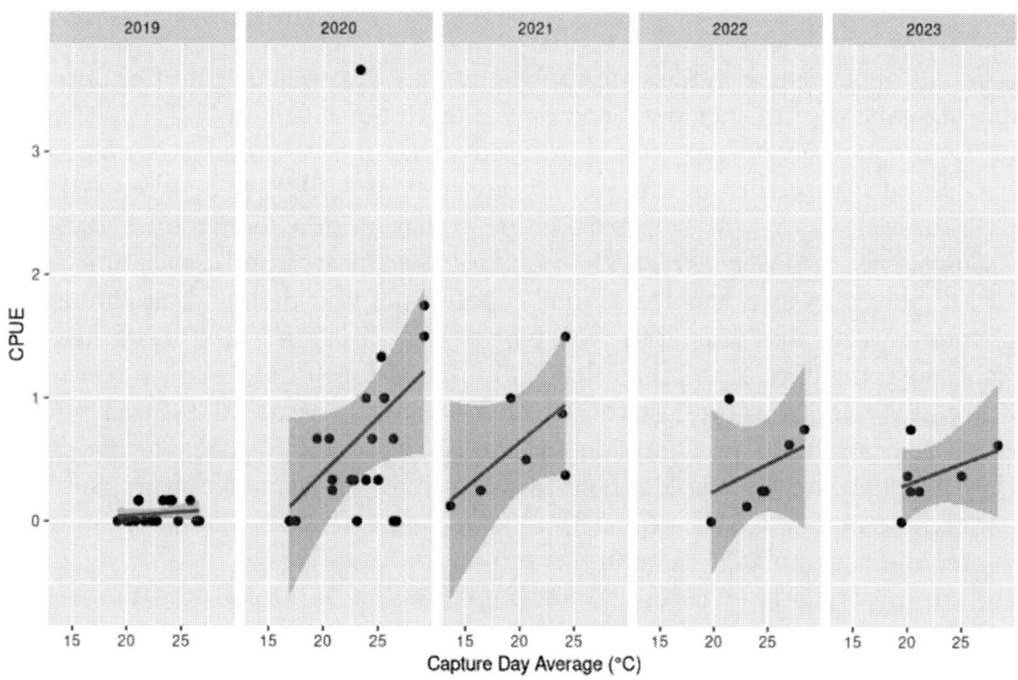

Figure 2. Linear relationship between daily CPUE and the average temperature of the day of capture (P = 0.005). Note that the fitted curve (with confidence band) is for each year, but the model discussed in the paper uses only one global slope across all years.

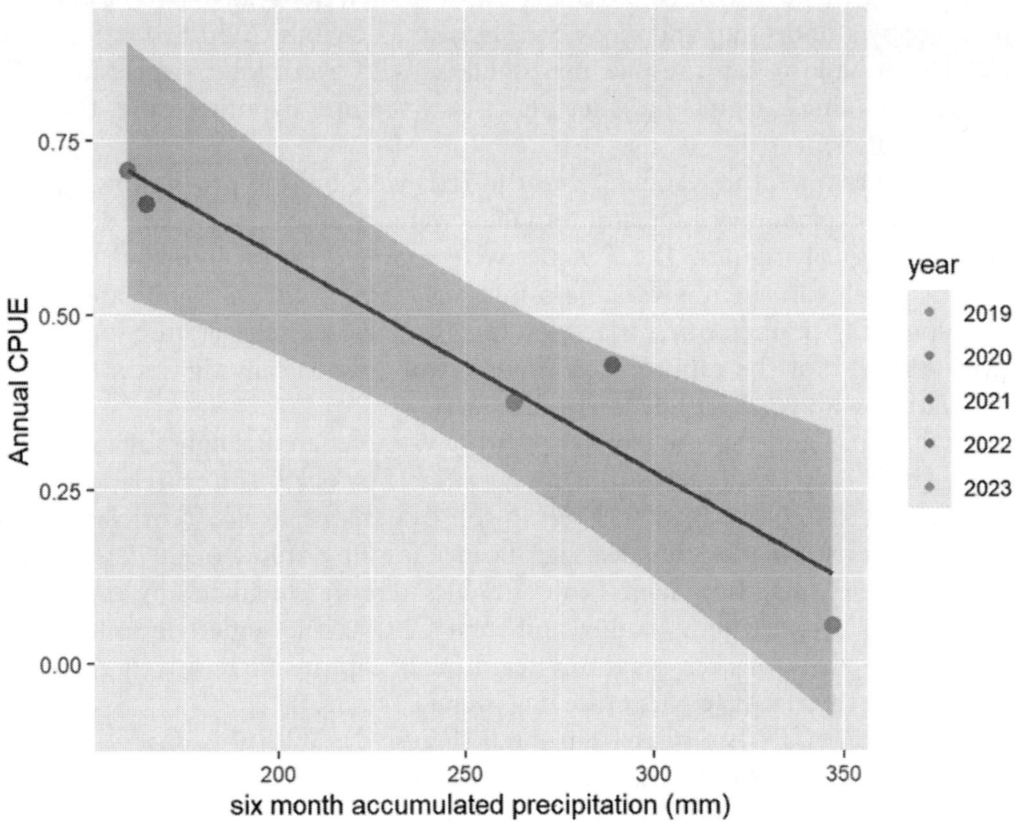

Figure 3. Line of best fit with confidence bands. Total CPUE and total precipitation over the previous six months.

residuals of two linear models are consistent with a normal distribution: previous day high and capture day low (Shapiro Wilkes test of normality, $W = 0.9668$, $P = 0.087$ and $W = 0.9727$, $P = 0.175$, respectively). All of the models, with or without the outlier, predict an average increase in daily CPUE of between 0.03 and 0.06 for each increase of 1 degree Celsius in that model's temperature variables.

Long-term weather variables—For longer-term weather associations, total CPUE per year was calculated as total captures per year divided by total number of trap days for that year (Table 1). Notice that 2019 and 2020 both had substantially more days of trapping than 2021, 2022, or 2023.

We used data from the Sisseton weather station for years 2019-2023. We calculated accumulated precipitation for the 12 months (Jul-Jun), the six month (Jan-Jun) and for the winters (Dec-Mar) prior to the start of capture and compared these to total CPUE (Table 3). Due to their extensive overlap, accumulated winter and accumulated six-month precipitation values show a highly significant and strong correlation with each other ($r = 0.94$, df = 3, P-value = 0.018). Twelve-month accumulation and six-month accumulation do not ($r = 0.52$, df = 3, P-value = 0.37) nor do twelve-month and winter ($r = 0.40$, df = 3, P-value = 0.51).

There is a statistically significant, negative correlation between annual CPUE and six-month accumulated precipitation ($r = -0.96$, df = 3, P-value = 0.009) (Figure 3), as well as a statistically significant, negative correlation between annual CPUE and winter accumulated rainfall (r = -0.89, df = 3, P-value = 0.04). There is not a statistically significant correlation between annual CPUE and twelve-month accumulated precipitation (r = -0.35, df = 3, P-value = 0.56).

The linear model has an adjusted r-squared value of 0.897 and all of the coefficients are statistically significant (Table 4). On average we see a decrease of annual CPUE of 0.003 animals per trap day for every increase in accumulated precipitation of 1 mm.

Table 1. Number of turtles captured in basking traps and hoops traps, total number of turtles captured, and calculation of catch per unit effort (in trap-days).

	2019	2020	2021	2022	2023
Basking Captures	1	24	2	3	1
Hoop Captures	6	27	35	21	20
Total Captures	7	51	37	24	21
Capture Days	20	23	7	7	7
Total Trap Days	128	72	56	56	56
Total CPUE	0.055	0.708	0.661	0.429	0.375

Table 2. *P*-values for the coefficients of each linear regression. Each row represents a separate linear model. All linear models have the same four indicator variables (2020, 2021, 2022, 2023). These establish a baseline for the CPUE for that year. The values in the column "Row variable" correspond to the *P*-value for the temperature variable unique to that model, indicated in the first column. Significant *P*-values are highlighted.

	Intercept	2020	2021	2022	2023	Row variable
7-day temp high	0.740	**0.004**	0.110	0.406	0.242	0.714
7-day temp low	0.874	**0.000**	**0.020**	0.156	0.254	0.908
7-day temp avg	0.926	**0.001**	**0.025**	0.253	0.196	0.896
3-day temp high	0.417	**0.001**	**0.034**	0.215	0.353	0.365
3-day temp low	0.265	**0.000**	**0.008**	0.163	0.172	0.225
3-day temp avg	0.333	**0.001**	**0.017**	0.197	0.273	0.288
Prev. day high	**0.018**	**0.002**	**0.039**	0.513	0.298	**0.015**
Prev day low	**0.031**	**0.000**	**0.002**	0.147	0.053	**0.024**
Prev day avg	**0.010**	**0.001**	**0.007**	0.363	0.113	**0.008**
Cap day high	**0.006**	**0.002**	**0.013**	0.416	0.356	**0.004**
Cap day low	**0.039**	**0.000**	**0.002**	0.116	0.070	**0.030**
Cap day avg	**0.006**	**0.001**	**0.002**	0.252	0.140	**0.005**

Table 3. Total precipitation (in mm) and catch per unit effort (CPUE = turtles per trap-day).

Year	6-Months	12-Months	Winter	CPUE
2019	346.71	724.66	133.86	0.055
2020	160.27	703.58	48.77	0.708
2021	165.35	350.27	47.50	0.661
2022	288.80	859.54	114.05	0.429
2023	262.89	509.78	129.54	0.375

Table 4. Coefficients, standard errors, t test statistic and *P*-value for the coefficients produced by the linear regression relating annual CPUE to six-month accumulated precipitation.

	Estimate	Std. Error	t value	Pr (>|t|)
Intercept	1.206	0.132	9.106	0.003
Six-month precipitation	-0.003	0.001	-5.987	0.009

DISCUSSION

We set out to examine the influence of short- and long-term weather variables on our ability to capture painted turtles and found effects of both. Accurate population surveys are crucial for tracking populations over time, and single-sample studies risk generating wildly misleading population estimates (Zweifel 1989). As mentioned earlier, the number of turtles captured during a survey can be influenced by capture methods, by season (McAuliffe 1978), and potentially by short-term and long-term environmental conditions (e.g. Moos and Blackwell 2016). Thus, long-term population studies are needed to develop an understanding of how external factors will influence the outcome of population studies.

Short-term local temperatures influenced our ability to capture turtles such that we captured more turtles on warmer days or the day following a warmer day. This same pattern holds for all models examined whether they satisfied the normality conditions of the residuals or not. Because all the models have similar coefficients, we are confident the data show increased short-term temperature values which result in increased average daily CPUE for our study site. We believe this is due to a positive association between temperature and overall turtle activity. As ectotherms, turtles rely on environmental temperatures for thermoregulation. Increased ambient temperatures have been shown to increase metabolic rate (Haskins and Tuberville 2022). Increased energetic expenditure presumably requires an increased need for energetic input, leading to an increase in foraging behavior; indeed, environmental temperatures have been shown to affect activity levels and time budgets in turtles (Marchant et al. 2021). Therefore, when environmental temperatures are warmer, the turtles move more and are more likely to encounter one of our traps.

The relationship between captures and local temperature does not hold when we examine the temperatures over slightly longer time frames (three-day and seven-day temperature averages). This is to be expected if the immediate, local temperatures directly determine turtles' activity levels. Grayson and Dorcas (2004) found that shell temperatures of painted turtles were strongly correlated with water temperatures. Again, as turtles are ectothermic, it seems likely that their body temperatures, and therefore activity levels, will be more dependent on the current temperature than the temperature on a previous day. The effects of environmental temperature on other aspects of turtle life history, such as nesting and egg-laying (Hedrick et al. 2021), incubation and hatching phenology, hatching success and emergence (Santoro 2023), and offspring sex ratios (Powell et al. 2023), have been well documented.

We found a strong negative correlation between precipitation in the previous 6-month period and total CPUE. For every 1 mm increase in rainfall during the previous six months, we captured 0.003 less animals per trap day. Put simply, when conditions were dry, we captured more turtles; when conditions were wet, we captured fewer turtles. This pattern was also observed in a study involving turtle bycatch in lakes and impoundments in northeastern South Dakota (Moos and Blackwell 2016). It is highly improbable that wet conditions negatively affected the size of our study population. Turtles, including painted turtles, are slow to reach sexual maturity, and therefore have long generation times (Ernst and Lovich 2009). A mortality event during a wet year that resulted in an actual decline in number of turtles would be detectable for many years after, while our CPUE was highest the two years following our lowest CPUE (see Table 3 and Figure 3). We might also expect to see a resultant shift in population structure, with hatchlings and juveniles more strongly represented in subsequent years. In fact, many studies have shown that drought or drought-like conditions due to wetland draw-downs result in increased freshwater turtle mortality [e.g., European pond turtles (*Emys orbicularis*) (Owen-Jones et al. 2015); Blanding's Turtles (*Emydoidea blandingii*) (Hall and Cuthburt 2000); Western Pond Turtle (*Actinemys marmorata*) (Purcell et al. 2017); Western Painted Turtles (Lindeman and Rabe 1990 and Beard and Powell 2023)]. A better explanation could be that drier conditions in our study area decreased available habitat, concentrating turtles, while wet conditions increased the amount of suitable habitat. Moos and Blackwell (2016) proposed that a wetter spring allows or encourages more turtles to leave the hibernation pond for other summer habitats. For Midland Painted Turtle (*Chrysemys picta marginata*), total daily distance moved decreased when conditions were drier (Rowe 2003). The author proposed several hypotheses including that contracting habitat may have concentrated prey items leading to decreased need for foraging activity (Rowe 2003). McAuliffe (1978) observed emigration from deeper water sources into the surrounding shallow oxbow system when conditions were wet and immigration back into the deeper slough as outlying areas dried. Likewise, painted turtles occupying a complex of ponds in Virginia moved between ponds only once per year but increased in density in three main ponds as the other ponds dried. One pond with low quality habitat was occupied only when other ponds had dried, suggesting that the turtles were demonstrating habitat selection (Bowne et al 2006).

Our observations underscore the need for long-term monitoring of population dynamics to account for the effect of environmental factors on sampling efficacy. Our future work will build on this by further investigating the parameters of our study population. We also plan to determine the extent of habitat use by our turtle population through the use of mark-recapture and radiotelemetry. It will also be important to know the corridors of travel, as painted turtles can travel along streams or over land (MacCulloch and Secoy 1983; Rowe 2003). This will allow us to determine connectivity between local water bodies, in terms of turtle usage, and the degree to which roads and traffic may pose a barrier to movement. And finally, a better understanding of the relationship between seasonal precipitation and the way in which turtles use their landscape will be important in light of predicted changes in temperature and patterns of precipitation due to climate change (Johnson et al. 2005).

ACKNOWLEDGEMENTS

We would like to thank the following agencies for funding this research: South Dakota Game, Fish and Parks Small Grant Program, University of Minnesota Faculty Research Enhancement Fund, Northern State University Northern Edge Award and Sanford Faculty Development Award. Research was conducted under South Dakota State University Institutional Animal Care and Use Committee protocol #19-032A (2019-2021) and #2205-029A (2022-2023), and SDGFP Scientific Collector's permit #41 (2019), #37 (2020 and 2021), #34 (2022) and #30 (2023). Many thanks to Terrina Barnes, Emily Schumacher, Jess Malchow, Haley Kastigar, Mya Krause, Maddie Gonzales and Mary Fites for their invaluable assistance on this project.

LITERATURE CITED

Beard, A.N., and L.A. Powell. 2023. Drought influences annual survival of Painted Turtles in Western Nebraska. Diversity 15(5):597. https://doi.org/10.3390/d15050597.

Bowne, D.R., M.A. Bowers and J.E. Hines. 2006. Connectivity in an agricultural landscape as reflected by interpond movements of a freshwater turtle. Conservation Biology 20: 780-791. https://doi.org/10.1111/j.1523-1739.2006.00355.x.

Davis, D.R. 2023. Painted Turtle. Amphibians and Reptiles of South Dakota. https://www.sdherps.org/species/chrysemys_picta. [accessed 30 June 2024].

dos Santos, R.L., J.M. de Sousa Correia, and E.M. dos Santos. 2021. Freshwater aquatic reptiles (Testudines and Crocodylia) as biomonitor models in assessing environmental contamination by inorganic elements and the main analytical techniques used: a review. Environmental Monitoring and Assessment 193:498. https://doi.org/10.1007/s10661-021-09212-w

Ennen, J.R., K.K. Cecala, P. Gould, R. Colvin, J. Denison, D.F. Garig, S. Hyder, L. Recker, and J.M. Davenport. 2021. Size matters: the influence of trap and mesh size on turtle captures. Wildlife Society Bulletin 45: 130-137. https://doi.org/10.1002/wsb.1163

Ernst, C.H., and J.E. Lovich. 2009. *Chrysemys picta*. Pages 184-211 *in* Turtles of the United States and Canada, 2nd ed. Johns Hopkins University Press, Baltimore, MD.

Gamble, T. 2006. The relative efficiency of basking and hoop traps for Painted Turtles (*Chrysemys picta*). Herpetological Review 37: 308–312.

Grayson, K.L., and M.E. Dorcas. 2004. Seasonal temperature variation in the Painted Turtle (*Chrysemys picta*). Herpetologica 60: 325-336.

Haskins, D.L., and T.D. Tuberville. 2022. Metabolic responses to increased temperatures in three semi-aquatic turtle species from the southeastern United States, Journal of Thermal Biology 109:103331. https://doi.org/10.1016/j.jtherbio.2022.103331.

Hall, C.D., and F.J. Cuthbert. 2000. Impact of a controlled wetland drawdown on Blanding's turtles in Minnesota. Chelonian Conservation Biology 3: 643–649.

Hedrick, A.R., D.U. Greene, E.L. Lewis, A.S. Hood, J.B. Iverson. 2021. Climate effects on nesting phenology in Nebraska turtles. Ecology and Evolution 11: 1225–1239. https://doi.org/10.1002/ece3.7105

Helms, T. D. 2010. Facility plan for a wastewater collection and treatment system for the Clear Lake Betterment Association, Marshall County, South Dakota. https://clbasd.com/wp-content/uploads/2022/05/clba_wastewater_facility_plan.pdf [accessed 29 April 2023].

Hollender, E., D. Ligon, and D. McKnight. 2022. Learned avoidance of trap locations in freshwater turtles. Wildlife Research 50: 10.1071/WR21061.

Jackson, D.C. 2002. Hibernating without oxygen: physiological adaptations of the Painted Turtle. The Journal of Physiology 543:731-737. https://doi.org/10.1113/jphysiol.2002.024729

Johnson, W.C., B.V. Millett, T. Gilmanov, R.A. Voldseth, G.R. Guntenspergen, and D.E. Naugle. 2005. Vulnerability of northern prairie wetlands to climate change. BioScience 55(10): 863-872.

Kornilev, Y.V., C.K. Dodd, and G.R. Johnston. 2012. Retention of paint markings for individual identification of free-ranging basking aquatic turtles (Suwannee Cooters, *Pseudemys concinna suwanniensis*). Herpetological Review 43: 61-64.

Krebs, C.J. 1999. Ecological Methodology, 2nd ed. Benjamin Cummings, Menlo Park, CA.

Lindeman, P.V., and F.W. Rabe. 1990. Effect of drought on the western painted turtle, *Chrysemys picta belli*, in a small wetland ecosystem. Journal of Freshwater Ecology 5: 359–364. https://doi.org/10.1080/02705060.1990.9665248

Lovich, J.E., J.P. Ennen, M. Agha, and J.W. Gibbons. 2018. Where have all the turtles gone, and why does it matter? BioScience 68: 771-781. https://doi.org/10.1093/biosci/biy095

Mali, I., D.J. Brown, J.R. Ferrato, and M.R.J. Forstner. 2014. Sampling freshwater turtle populations using hoop nets: testing potential biases. Wildlife Society Bulletin 38: 580–585.

Mali, I., D.J. Brown, M.C. Jones, and M.R.J. Forstner. 2012. Switching bait as a method to improve freshwater turtle capture and recapture success with hoop net traps. Southeastern Naturalist 11: 311–318.

MacCulloch, R.D., and D.M. Secoy. 1983. Demography, growth, and food of western painted turtles, *Chrysemys picta bellii* (Gray), from southern Saskatchewan. Canadian Journal of Zoology 61: 1499–1509.

Marchand T., A.-S. Le Gal, and J.-Y. Georges. 2021. Fine scale behaviour and time-budget in the cryptic ectotherm European pond turtle *Emys orbicularis*. PLoS One 16(10): e0256549. doi: 10.1371/journal.pone.0256549.

McAuliffe, J. 1978. Seasonal migrational movements of a population of the western painted turtle, *Chrysemys picta bellii* (Reptilia, Testudines, Testudinidae). Journal of Herpetology 12(2): 143-149.

Moos, T.S., and B.G. Blackwell. 2016. Characterization of western painted turtle bycatch in fyke nets during freshwater fish population assessments. Journal of Fish and Wildlife Management 7: 222-230. doi: 10.3996/102014-JFWM-077

Moos, T.S. and B.G. Blackwell. 2017. Condition and mortality of western painted turtles collected as bycatch in modified fyke nets during fish population assessments. Journal of Freshwater Ecology 32: 267-280. doi: 10.1080/02705060.2017.1281174

Owen-Jones, Z., P. Priol, S. Thienpont, M. Cheylan, G. Sauret, C. Coïc, and A. Besnard. 2015. The contrasting effects of short- and long-term habitat drainage on the population dynamics of freshwater turtles in a human-dominated landscape. Freshwater Biology 61: 121-132. https://doi.org/10.1111/fwb.12687.

Powell, L.A., E.P. Dolph, and C.R. Neil. 2023. Drought affects sex ratio and growth of painted turtles in a long-term study in Nebraska. Climate Change Ecology 5: 100072. https://doi.org/10.1016/j.ecochg.2023.100072.

Purcell K.L., E.L. McGregor, and K. Calderala. 2017. Effects of drought on western pond turtle survival and movement patterns. Journal of Fish and Wildlife Management 8: 15– 27.

Rowe, J.W. 2003. Activity and movements of midland painted turtles (*Chrysemys picta marginata*) living in a small marsh system on Beaver Island, Michigan. Journal of Herpetology 37: 342-353. https://doi.org/10.1670/0022-1511.

Santoro, A., J.M. Chambers, B. Ebner and S.J. Beatty. 2023. Climate change affects the early-life history of a freshwater turtle in a severely drying region. Austral Ecology 48: 2205-2229. https://doi.org/10.1111/aec.13456

Stanford, C.B., J. B. Iverson, A.G.J. Rhodin, P.P. van Dijk, R.A. Mittermeier, G. Kuchling, K.H. Berry, A. Bertolero, K.A. Bjorndal, T.E.G. Blanck, K.A. Buhlmann, R.L. Burke, J.D. Congdon, T. Diagne, T. Edwards, C.C. Eisemberg, J.R. Ennen, G. Forero-Medina, M. Frankel, U. Fritz, N. Gallego-García, A. Georges, J.W. Gibbons, S. Gong, E.V. Goode, H.T. Shi, H. Hoang, M.D. Hofmeyr, B.D. Horne, R. Hudson, J.O. Juvik, R.A. Kiester, P. Koval, M. Le, P.V. Lindeman, J.E. Lovich, L. Luiselli, T.E.M. McCormack, G.A. Meyer, V.P. Páez, K. Platt, S.G. Platt, P.C.H. Pritchard, H.R. Quinn, W.M. Roosenburg, J.A. Seminoff, H. B. Shaffer, R. Spencer, J.U. Van Dyke, R.C. Vogt, and A.D. Walde. 2020. Turtles and tortoises are in trouble. Current Biology 30(12): R721-R735. https://doi.org/10.1016/j.cub.2020.04.088.

Tesche, M.R. and K.E. Hodges. 2015. Unreliable population inferences from common trapping practices for freshwater turtles. Global Ecology and Conservation 3: 802-813. http://dx.doi.org/10.1016/j.gecco.2015.04.001

Waye H., A. Dolan, and P. Dolan. 2023. Emerging patterns in population structure and trap efficacy after three years of a survey of western painted turtles (*Chrysemys picta bellii*, Gray, 1830) in Marshall County, South Dakota. Prairie Naturalist 55: 124–134.

Zweifel, R.G. 1989. Long-term ecological studies on a population of painted turtles, *Chrysemys picta*, on Long Island, New York. American Museum Novitates No. 2952, 55 pp.

Abstracts of Research Papers and Posters

submitted for presentation at the

109th Annual Meeting

of the

South Dakota Academy of Science

WHICH IS THE BEST: TECHNOLOGY, RESOURCES OR ACTIVE LEARNING ACTIVITIES?

Pankaj Mehrotra*
University of the People
Pasadena, CA, 91101
*Corresponding author email: mehrotra85pankaj@gmail.com; pankaj.mehrotra@uopeople.edu

ABSTRACT

Science Learning Spaces (SLS) are zones of interaction among undergraduates and between undergraduates and professors. Science Learning Spaces have evolved from traditional learning spaces to multidisciplinary and interdisciplinary active learning science spaces. The use of technology, providing additional resources and both individual and collaborative active learning activities enhances students learning of science apart from traditional discussion-based learning. The interaction in science learning spaces enhances student ability to learn content and attain skills through retention and development of student identity and attainment of leadership roles. Learning spaces are of two types: Major Learning Spaces and Minor Learning Spaces. In my previous oral presentation, I presented the Major Learning Spaces and Minor Learning Spaces in in-person and Virtual Learning Environment. This presentation will be in continuation of the previous presentation on use of technology, such as Jamboard, Google doc, Google sheet, and how they enhanced science learning in major and minor learning spaces. Further, I will show how science students perform in the major and minor learning spaces when they participate in individual and group learning activities. Additionally, sharing of resources such as research articles and videos in learning spaces further enhances science learning. I will be presenting my original work describing the impact on science learning in learning spaces when technology, additional resources and appropriate learning activities are integrated. The focus of the presentation will also be on sharing information regarding what are the better ways to organize and build student groups in STEM courses. Overall thoughtful selection of technology, appropriate resources and selection of active learning activities enhances student learning outcomes in learning spaces.

EVALUATION OF ALTSHULER LINEAR ANTENNAS PRODUCED USING PRINTING TECHNIQUES

Thomas P. Montoya*
Department of Electrical Engineering & Computer Science
South Dakota School of Mines & Technology
501 E. Saint Joseph St
Rapid City, SD 57701-3995
Corresponding author e-mail: Thomas.Montoya@sdsmt.edu

ABSTRACT

Research results evaluating the performance of Altshuler linear antennas (E. E. Altshuler, ``The Traveling-Wave Linear Antenna," *IRE Trans. Antennas Propagation*, pp. 324-329, July 1961.) produced using printing techniques are presented. These antennas are conductive dipoles/monopoles with discrete resistive loading placed a quarter-wavelength from the open end(s). Altshuler linear antennas typically have input impedance bandwidths of about 2 to 1 (ratio of the highest to lowest acceptable frequency) while maintaining broadside omnidirectional radiation patterns. Comparable conductive halfwave dipoles and quarterwave monopoles (no resistive loading) have bandwidths of a few percent about the resonant frequency and while the radiation patterns shift from broadside above this frequency. The conductive arms (traces) of the experimental Altshuler monopoles were produced by depositing commercially-available materials using an nScrypt system onto a flexible polyimide substrate. The resistive loading was realized by soldering surface mount chip resistors across gaps in the traces. The measured input impedances as well as elevation radiation patterns were compared to simulated results. The simulated results were obtained using Numerical Electromagnetics Code (NEC 2.0) models. The experimental and simulated impedance results are in good agreement. The radiation pattern results, while validating the expected broadside main lobe, showed more discrepancies. In summary, this research demonstrated the successful construction of an Altshuler monopole using printing techniques.

GENETIC RESOURCES FOR ENHANCED ALFALFA PERSISTENCE: UTILIZING CROWN AND ROOT SPROUTING TRAITS IN GERMPLASM EVALUATION

Abdulallah M. Alomair[1], Arvid Boe[2], and Lan Xu[1]*
[1]Department of Natural Resource Management
[2]Department of Agronomy Horticulture & Plant Science
South Dakota State University
Brookings, SD 57006
*Corresponding Author: Lan.Xu@sdstate.edu

ABSTRACT

Alfalfa ranks as the fourth largest crop in the U.S. and is the only perennial among crop species. Its production plays a crucial role in sustaining the beef and dairy industries. However, alfalfa faces challenges in persisting in semi-arid and arid rangelands. Research indicates that alfalfa plants with deep crowns and fibrous root systems exhibit greater tolerance to drought, grazing, and winter hardiness conditions, thereby ensuring better persistence. This study aimed to screen, identify, and evaluate alfalfa germplasm for crown and root morphological traits related to alfalfa persistence in rangelands. Thirteen yellow-flowered alfalfa (*Medicago falcata*) accessions from the National Plant Germplasm System (NPGS) and one commercial cultivar (Final Answer, *Medicago sativa*) as the control were evaluated, with trials repeated twice in 2021 and 2022. One hundred uniform seeds were selected from each germplasm, scarified, and inoculated with Rhizobium before planting in plastic conical containers in a greenhouse. After the first year of dormancy, all plants were excavated to examine the crown traits, including the crown diameter, number of crown stems, and crown bud density. Additionally, primary roots were cut into two six-centimeter-long segments at one and seven centimeters below the cotyledonary node to assess sprouting ability. We found that crown bud density and root sprouting ability, which are often considered the most critical traits for alfalfa persistence, varied among the selected germplasms. The control group, Final Answer, exhibited one of lowest crown bud densities with no root sprouting capacity, while PI 634114 not only demonstrated a high crown bud density but also a notable frequency of root sprouting. Particularly noteworthy were multiple germplasms originally from Kazakhstan displaying these desirable traits, suggesting they can potentially serve as valuable genetic resources for breeding programs aiming to enhance alfalfa persistence.

EDUCATION ON THE UNDERSTANDING OF NOISE-INDUCED HEARING LOSS IN VULNERABLE UNDERGRADUATE POPULATION

Henry Wei* and Andrew Ridder
Sanford School of Medicine
University of South Dakota
Sioux Falls, SD 57105
*Corresponding author: Henry.Wei@coyotes.usd.edu

ABSTRACT

With the introduction of increasingly powerful audio equipment and increase of personal mobile audio devices in the 21st century, the prevalence of noise-induced hearing loss (NIHL) in young adults is expected to increase. This increase, estimated to impact 30 million adults in the next four decades, is due in part to recreational exposure. While many young adults have a general understanding of NIHL, a detailed education on various topics of NIHL could further promote adherence to the use of preventive measures. This project therefore involved gauging understanding of NIHL and providing in-depth education on various topics of NIHL geared to an undergraduate population at 3 universities located in South Dakota.

The project was approved by the Institutional Review Board (IRB) of the University of South Dakota. A forty-five-minute PowerPoint presentation was developed and presented about NIHL. The following 5-topic areas were discussed: pathophysiology of hearing loss, healthcare impacts, Occupational Safety and Health Administration (OSHA) guidelines on noise exposure, adherence to OSHA guidelines in recreational spaces, and preventive measures, to include decibel meters and earplugs. PowerPoint presentations and pre- post-surveys were given during the recital hour, a required course, to 120 students enrolled in music degree programs at 3 universities in South Dakota. A pre-presentation survey was given to score students' baseline knowledge in each of the topic areas covered, followed by a post-presentation survey to score knowledge gained. A one-tailed paired t-test was performed to evaluate the changes from pre- to post surveys. In addition to the education, decibel meter mobile applications were introduced, and 120 high-fidelity earplugs were donated at the sites of presentation.

Comparison of all the pre- to post-survey responder scores showed a statistically significant change in the students' confidence in their knowledge for all 5-topic areas discussed. There was also statistically significant change in knowledge confidence in all 5-topic areas between responder groups who were already using preventive measures compared to those who were not using preventive measures at the time of testing. For example, the knowledge confidence to wear preventive measures notably increased nearly threefold, 1.68 to 4.26 amongst those who were not using preventive measures.

This project demonstrated that young adults with a current knowledge of NIHL still gained additional insight through this presentation. While there was a significant change in knowledge confidence in each category in the total responders, the project also demonstrated that amongst current users of preventive measures, there was a statistically significant change in knowledge confidence as well. Students also indicated, through the surveys, that they are more likely to use preventive measures in loud recreational spaces after this educational experience. Many but not all the students took advantage of the high-fidelity earplug donations.

UNRAVELING EASTER REDCEDAR ENCROACHMENT IN GRASSLANDS THROUGH SEED VIABILITY OVER TIME

Julienne Korst[1], Arvid Boe[2] and Lan Xu[1]*
[1]Department of Natural Resource Management
[2]Department of Agronomy Horticulture & Plant Science
South Dakota State University
Brookings, SD 57006
*Corresponding Author: Lan.Xu@sdstate.edu

ABSTRACT

Woody encroachment has emerged as a global threat to grassland biomes and the people, plants and animals that depend on them. Eastern red cedar (ERC) (*Juniperus virginiana*) recruitment and expansion are driven mainly by its prolific seed production and seed dispersal via birds and small mammals. However, our understanding of ERC soil seed bank dynamics, including seed bank size, seed viability, and longevity remains scarce. The objective of this study was to evaluate ERC seed viability in the soil seed bank across different vegetation types over time. The study site was located in the eastern South Dakota grasslands and consisted of four blocks, half of them dominated by warm-season native vegetation and the other two dominated by cool-season introduced grass species. To assess seed viability over time in the field, we randomly placed five seed mesh bags, each containing 25 intact seeds collected from fully ripened and intact ERC fruit within each block in contact with the soil surface. After predetermined intervals of 0 (as a control), 34, 166, 246, and 480 days, the bags were randomly retrieved to evaluate seed viability using TZ Procedure. Surprisingly, we found that ERC seed viability consistently exceeded 50% across all treatments over 16 months study duration. Furthermore, statistical analyses revealed no significant difference in seed viability among the treatments within the same sampling dates or over time within the same treatment. These findings suggest that ERC seed viability remains robust and resilient, unaffected by types of grasslands dominated by cool-season or warm-season grasses over the course of 16 months.

AN UNUSUAL NEW INSECT GALL ON *MUHLENBERGIA CUSPIDATA* (POALES: POACEAE)

Paul J. Johnson* and Arvid Boe
Department of Agronomy, Horticulture, and Plant Science
South Dakota State University
Brookings, South Dakota 57007
*Corresponding author email: conoderus@ gmail.com

ABSTRACT

Plains muhly *(Muhlenbergia cuspidata* (Torr. ex Hook.) Rydb. (Poales: Poaceae)), sometimes also known as Stoney Hills muhly, is a native North American prairie grass with historical records from Pennsylvania and Virginia to Alberta and New Mexico. It is moderately palatable to livestock and is removed under regular grazing. In eastern South Dakota and western Minnesota, we normally find it on remnant ungrazed slopes of well-drained glacial deposits and steep loess, and roadcuts, and in Western South Dakota, North Dakota, and Montana on well-drained slopes of buttes and canyons. We investigated populations of the plant from 2019–2023 at 10 sites in Codington, Deuel, Grant, and Roberts counties, South Dakota. At most populations we frequently found an abnormal spikelet structure that revealed itself as an insect induced floret gall (Figure 1), likely the same as noted from North Dakota by O.A. Stevens in 1963. These galls are identical in structure to ones found previously by us in northeastern Montana. The same structures were also found on *M. cuspidata* specimen images in online herbaria from sites in Wyoming and New Mexico. In some of the botanical literature this gall appears to have been confused with the sori of the smut *Tilletia asperifolioides* G.W. Fischer (Basidiomycota: Tilletiales: Tilletiaceae). This muhly gall is shining pale yellow-green to yellow-tan colored, is globular basally, and attenuates apically, which gives a beak-like appearance from exaggerated growth of the palea that has its margins rolled, fused basally, and overlapping apically. The gall is hollow and the apical extension forms a narrow tube. Dissections of immature galls revealed the present of translucent white to pale-red colored larva of a gall midge (Diptera: Cecidomyiidae) of an undescribed species attributable to the subfamily Lasiopterinae, possibly a *Dasineura Rhondani* species or closely related (R. Gagné, *in litt.*). These larvae were feeding on the ovary. Attempts at rearing adults of this fly have yet to provide adequate material for a formal taxonomic description. We also reared a parasitoid wasp attributable to the genus *Aprostocetus* Westwood (Hymenoptera: Eulophidae), that is also undescribed. Based on rearing and insect residue in late season galls it appears that the mature gall midge larva exits the gall in autumn through the tubular beak and overwinters in soil beneath the plants, where it pupates probably in late spring. The adult parasitoid emerges by chewing through the base of the gall in late summer and autumn to overwinter off the host plant. We observed that elevation of the growing point of reproductive tillers during stem elongation can result in destruction of the inflorescence by grazing, trampling, haying and mechanized crushing. In late spring and early summer, drought can prevent floral development. This seasonal absence of flowers appears to depress gall midge and parasitoid populations.

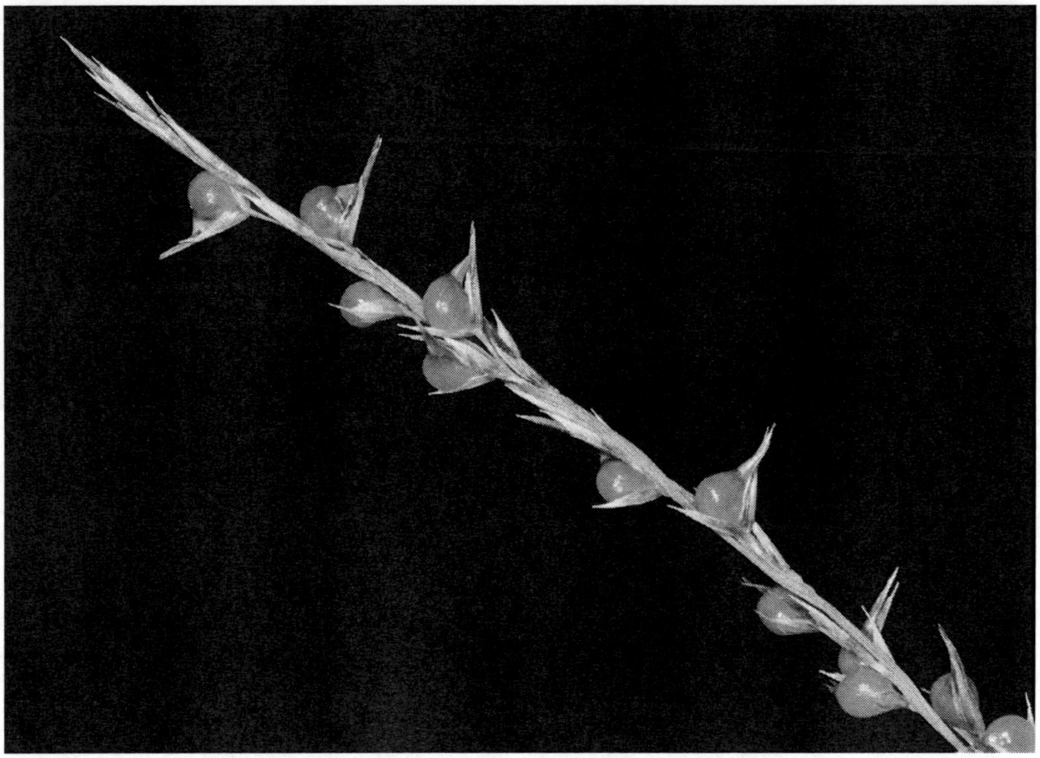

Figure 1. Apical half of a *Muhlenbergia cuspidata* inflorescence showing galls, each resulting from floret parasitism by a gall midge larva.

IMPACT OF EXOTIC FLOWERING PLANTS ON POLLINATOR VISITATION TO NATIVE SPECIES

Amadeu dos Santos Neto* and Charles B. Fenster
Department of Biology & Microbiology
Oak Lake Field Station
South Dakota State University
Brookings, SD 57006
*Corresponding author email: amadeu.dossantosneto@sdstate.edu

ABSTRACT

Invasive plant species represent a significant threat to biodiversity and agriculture, outcompeting and potentially excluding native species. Here, our objective is to understand the impact of the presence of exotic species on the visitation of pollinators to native species. Our study was conducted at the Oak Lake Field Station, Brookings County, South Dakota. We sampled 451 plants, belonging to 60 native species. We observed visits to 131 hosts plants (29%); small flies were the most important group (44%), followed by bees (39%), beetles (35%), small bees (20%), flies (16%), wasps (10%) and Lepidoptera (3%). We applied a logistic regression that includes "visitation" as the dependent variable, floral density of native species as the independent variable and the floral density of exotic species as a predictor variable. Other models were produced replacing visitation with numbers of specific groups (bees, small bees, flies, small flies, Lepidoptera, beetles and wasps). In the general model, we found a positive (0.28) and significant coefficient for exotic (z: 5.174, $p < 0.01$). In the model performed with visitation of bees, we found a negative (-0.24186) and significant coefficient with floral density of exotics (z: -2.118, $p < 0.05$). For small flies and beetles, we have positive (0.20 and 0.21, respectively) and significant coefficients with exotic species (small flies - z: 3.101, $p < 0.01$, beetles - z: 3.058, $p < 0.01$). Our results suggest that the floral density of exotic species surrounding native species can increase the chances of visitation. However, for larger bees (which includes *Bombus* spp.), the chances of visiting apparently decrease with the increase in the floral density of exotic species. The increase in the density of exotic species apparently increases the chances of observing more generalist species (such as small flies and bees).

REACT! GAMIFYING ORGANIC REACTION LEARNING IN NON-MAJORS CHEMISTRY CLASSES

Paula Mazzer
Chemistry Department
Dakota Wesleyan University
Mitchell, SD 57301
Corresponding author email: paula.mazzer@dwu.edu

ABSTRACT

Students of organic chemistry, especially in non-majors courses, routinely struggle to predict products of organic reactions. To encourage learning, I developed a game to teach the principles of product prediction to non-majors chemistry classes. Game play involves drawing reaction cards to use to complete a three-step reaction pathway. Reaction cards describe in words the product of their reaction, and a table at the bottom of the card highlights which starting materials can undergo that type of reaction. Starting from a central, randomly selected starting material, each competitor attempts to complete their own pathway by playing a reaction card and sketching the product—which then becomes the reactant of the subsequent step. Playing the game in a non-majors general, organic, and biochemistry course this spring dramatically improved scores on product prediction questions on the exam. Further development of the game will involve creating a modification to make it more useful to majors-level organic synthesis courses.

INSTRUCTIONS FOR AUTHORS

Editorial Policy: The South Dakota Academy of Science is a forum for the promotion of scientific teaching, research, and service in the state of South Dakota. The South Dakota Academy of Science meets each spring for the purpose of annual business, awards, and the interchange of ideas, information, and results from scientists and students of science. The minutes of the annual meeting and the abstracts of presentations are published annually in the Proceedings of the South Dakota Academy of Science. In many cases, authors choose to submit a full manuscript of their presentation, which is published in lieu of the abstract. **Deadlines for submission of full manuscripts are July 15** each year; to submit manuscripts, please adhere to the Submission Procedures listed below. Authors have the option of purchasing additional reprints; information on page and reprint charges will accompany the page proofs sent to the corresponding author. The following guidelines provide instructions for authors submitting manuscripts to the Proceedings. Be aware that these guidelines may vary from year to year, so please try to format your manuscript so that it meets the conventions described below.

Authorship Policy: The following is the policy on authorship of abstracts and manuscripts to be published in the Proceedings:

1) All papers published in the Proceedings of the South Dakota Academy of Science will include all authors that were shown on originally submitted abstracts or manuscripts.

2) Any request to change authors after a submission has been made to the Proceedings must be accompanied by a brief letter of explanation sent to the editor that includes the original signature of all authors (including authors removed from or added to the authorship list).

3) The deadline for requesting changes in authorship is 30 August of the year of submission.

Manuscript Submission Procedure: Manuscripts submitted for publication in Proceedings must be accompanied by **two** technical reviews. Authors are required to solicit reviewers that have expertise in the topic area of the manuscript. Co-authors of manuscripts are not considered reviewers. Each reviewer must provide their name, title, address, and phone number along with a brief review of the manuscript; to facilitate this process, the reviewers should use the Peer Review Form that can be downloaded from the Academy's web site (http://sdstate.edu/sdas/). In turn, authors must provide a brief reconciliation letter describing how reviewer comments were addressed along with a final, revised copy of the manuscript. The submission package must include the following: 1) two technical reviews (Peer Review Forms), 2) written reconciliation from authors, and 3) revised manuscript. All file names should begin with the corresponding author's last name. Failure to adhere to the submission procedure will result in manuscripts being returned to authors.

Submission of abstracts for publication in the Proceedings **do not need** to be peer-reviewed.

Abstracts or manuscripts may be submitted either electronically or by mail. Please do not submit abstracts or manuscripts both electronically and by mail.

Electronic submission:
Electronic submissions should be forwarded to **Robert Tatina at rotatina@dwu.edu**. Manuscripts should be saved as **MS Word files** (not PDFs) that include tables and/or figures. Reviewer comments and the author(s) reconciliation statement must be submitted as separate files (MS Word) along with the manuscript. Manuscript illustration files that have the extension .BMP, .GIF, .EPS, .JPG, .TIF, or .PCX are acceptable. IMPORTANT: **Begin all file names with the corresponding author's last name** (e.g., Chipps_ms.doc, Chipps_review1.doc, Chipps_review2.doc, Chipps_reconcil.doc).

Mail submission:
Manuscripts may be submitted by mail by including one hardcopy and one electronic copy (3.5" disk or CD) of each manuscript. Hardcopies of reviewer's comments and reconciliation statement must also be included. Each diskette or CD should be identified with authors' names and by the format used. Submit all information in MS Word. Illustration files that have the extension .BMP, .GIF, .EPS, .JPG, .TIF, or .PCX are acceptable. **Begin all file names with the corresponding author's last name** (e.g., Chipps_ms.doc).

Mail manuscript submissions to:

Robert Tatina, Proceedings Editor
308 Orchard Blvd.
Sawyer, MI 49125

Phone: 269-426-3744
E-mail: rotatina@dwu.edu.

Membership. The author or one co-author of each abstract or paper must be a current member of the South Dakota Academy of Science.

Web Site Address. http://www.sdaos.org

Format Conventions: All manuscripts and abstracts should be single-spaced throughout including tables and the literature cited section. All pages should be numbered serially in the upper right-hand corner. Use a 1-inch margin all around but do not justify the right margin or hyphenate words on the right margin (i.e., use word wrap). Avoid footnotes in both articles and abstracts. Use metric measurements unless English units are clearly more appropriate, in which case metric equivalents must appear in parentheses. Give scientific names for all species the first time the species' common name is used as follows: Culver's root (*Veronicastrum virginicum*). Write out numbers under 10, except in a series with numbers >10 or with measurements or percentages. Statistical terms and other measures should conform to the Council of Science Editors Scientific Style and Format (e.g., $t = 0.311$, df = 33, $P = 0.001$, note italics, capital P and spaces before and after =).

Manuscript Components:

TITLE PAGE
All manuscripts should include a title page that includes the author name(s), affiliation(s), current mailing address(es), and Corresponding author email. Here are the components and style of the title page:

Microsoft Word
Times New Roman, Font Size = 12
Margin: Left = 0, Right = 6, Paragraph Indent = 4
Title: All Caps, Bold, Centered
Authors: See example:

Alfred Jones[1*], William Smith[1], and Lawrence Porter[2]
[1]Biology Department
Cognation University
South Chicago, IN 44230
[2]Chemistry Department
Intuition College
Newell, SD 57330
*Corresponding author email: alfred.jones@cogun.edu

ABSTRACT

Include a brief but informative abstract. The abstract should be a single paragraph of 200-300 words that concisely summarizes the results and conclusions of the study. The abstract should be followed by a short list of keywords (< 5). For authors submitting only an abstract, you may include a few literature citations, but please use sparingly, and enclose them in parentheses within the text of the abstract.

Keywords

(Up five keywords)

INTRODUCTION

The introduction should include sufficient detail to establish the purpose and importance of the work being summarized. It should include pertinent literature related to the study question and end with a statement of the objectives/hypotheses being examined.

METHODS

The methods should include sufficient information for the reader to evaluate the data or repeat the study. It is better to be inclusive (i.e., overly explicit) when describing the methods used, but keep in mind that clarity and conciseness are important.

RESULTS

Results follow the methods section and should include a concise summary of your findings. Be careful not to 'discuss' implications of results; reserve these points for the Discussion section. Results that are presented in tables or figures should not be exhaustively discussed in the text. When statistics are presented they should be explicit (e.g. t = 34.07, df = 48, P = 0.003).

DISCUSSION

The common trait among good scientific papers is a well-written discussion section. The discussion section is where you synthesize what has been learned from the study and provide direction or generalities for future studies. The discussion section should not be merely a summary of relevant literature linked to restatements of your results. Rather, it should be an informative essay that advances what is known about the subject area. Be careful, however, not to extend the discussion beyond what can be supported by the data reported in the results.

LITERATURE CITED

All references cited in the text should be listed at the end of the manuscript under the Literature Cited heading. References should be in alphabetical order by the last name of the author(s). If different works by the same author(s) are referenced, references should be in chronological order. Authors cited in the text of abstracts or papers should be indicated in parentheses by last name and year of publication (e.g., Wade 1967; Bremer 1977). When a citation has more than two authors, the last name of the first author followed by et al. should be used (e.g., Seabloom et al. 1978). Where two or more papers by the same author(s) have appeared in one year, the style should be (Stewart 1967a, 1967b). Citations included in abstract only submissions should be complete and included within the text of the abstract.

Examples of Literature Cited

Anderson, R.J., and R.W. Schrier. 2000. Acute renal failure. Pages 1149-1160 *in* Braunwald, E., K.J. Isselbacher and R.G. Petersdorf, editors. Harrison's principles of internal medicine. 15th ed. McGraw-Hill, New York, NY.
Bremer, P.E. 1977. Pelican kill. Loon 49:240-241.
Gipson, P.S., I.K. Gipson, and J.A. Sealander. 1975. Reproductive biology of the bobcat (*Lynx rufus*).Journal of Mammalogy 56:605-612.
Green, J.C., and R.B. Janssen. 1975. Minnesota birds: where, when, and how many. University of Minnesota Press, Minneapolis, MN.
Midwest Regional Climate Center. 2002. Historical climate summaries. Available at <http://www.hprcc.unl.edu/> [Cited 12 May 2012]
Stewart, A.V., A. Joachimiak, and N. Ellison. 2009. Genomic and geographic origins of timothy (*Phleum* sp.) based upon ITS and chloroplast sequences. Pages71-81*in* T. Yamada and G. Spangenberg, editors. Molecular breeding of forage and turf. Springer, New York, NY.
Svihovec, L.K. 1967. A comparison study of the ecological distribution of small mammals in southwestern North Dakota. Thesis. University of North Dakota, Grand Forks, ND.

(For Internet citations, include as much information as possible so that readers could access the material cited. Also include "Available at" and then include the <URL>. Finally, add in square brackets the date you accessed the URL [Cited 6 June 2010].

Tables, Figures, Illustrations and Photographs. Tables should be double spaced throughout, without vertical lines, and with page numbers in the upper right-hand corner. If tables are more than one page in length, put "Table x. continued" on all subsequent pages. Each table should begin with a Table number (e.g., Table 1, Table 2, etc…) followed by a descriptive caption which is separated from the table headings and data by a horizontal line. Tables must be referenced in the text.

Figures should be done with black ink on firm quality paper or if computer generated, they should be of camera-ready quality (laser printer) with sharp lines suitable for reproduction. Photographs should be 5x7-inch, sharp, black and white glossy prints. When appropriate, a legend of symbols, etc., scale and a cardinal direction (e.g. North) should be indicated on figures of maps. On the back side, identify each figure or photograph lightly in pencil with the figure's number, author(s) name(s), and an indication of the top of the figure. Figure captions should be numbered (e.g., Figure x.) and typed on a separate sheet of paper at the end of the manuscript. Letters, symbols, and legends should be large and clear enough on all figures and photographs to be legible after copy reduction. Figures must be referenced in the text.

A note on deadlines: If you cannot meet the July 15th deadline for full papers, you must send the editor a note requesting an extension. You must also request a new due date because extensions are not automatically granted.

South Dakota Academy of Science
308 Orchard Blvd.
Sawyer MI 49125

ISBN 979-8-3313-0813-1

CURRAN ASSOCIATES INC.
proceedings.com

2nd International Symposium on Incoherent Light Sources

Enschede, Netherlands
9 – 12 April 1979